ZIML Math Competition Book

Division H 2016-2017

Areteem Institute

TITLES PUBLISHED BY ARETEEM PRESS

Cracking the High School Math Competitions (and Solutions Manual) - Covering AMC 10 & 12, ARML, and ZIML
Mathematical Wisdom in Everyday Life (and Solutions Manual) - From Common Core to Math Competitions
Geometry Problem Solving for Middle School (and Solutions Manual) - From Common Core to Math Competitions
Fun Math Problem Solving For Elementary School (and Solutions Manual)
ZIML Math Competition Book Division E 2016-2017
ZIML Math Competition Book Division M 2016-2017
ZIML Math Competition Book Division H 2016-2017
ZIML Math Competition Book Jr Varsity 2016-2017
ZIML Math Competition Book Varsity Division 2016-2017

COMING SOON

Fun Math Problem Solving For Elementary School Vol. 2 (and Solutions Manual)
Counting & Probability for Middle School (and Solutions Manual) - From Common Core to Math Competitions
Number Theory Problem Solving for Middle School (and Solutions Manual) - From Common Core to Math Competitions

The books are available in paperback and Kindle eBook formats. To order the books, visit

https://areteem.org/bookstore.

ZIML Math Competition Book Division H 2016-2017

Edited by   John Lensmire
              David Reynoso
              Kevin Wang
              Kelly Ren

Cover and chapter title photographs by Kelly Ren and Kevin Wang

ISBN: 1-944863-12-5
ISBN-13: 978-1-944863-12-8
First printing, May 2018.

Contents

Introduction

Each month during the school year, Areteem Institute hosts the online Zoom International Math League (ZIML) competitions. Students can compete in one of five divisions based on their age and mathematical level (details shown on Page 7).

This book contains the problems, answers, and full solutions from the nine ZIML Division H Competitions held during the 2016-2017 School Year. It is divided into three parts:

1. The complete Division H ZIML Competitions (20 questions per competition) from October 2016 to June 2017.
2. The solutions for each of the competitions, including detailed work and helpful tricks.
3. An appendix including the topics and knowledge points covered for Division H, a glossary including common mathematical terms, and answer keys for each of the competitions so students can easily check their work.

The questions found on the ZIML competitions are meant to test your problem solving skills and train you to apply the knowledge you know to many different applications. We hope you enjoy the problems!

About Zoom International Math League

The Zoom International Math League (ZIML) has a simple goal: provide a platform for students to build and share their passion for math and other STEM fields with students from around the globe. Started in 2008 as the Southern California Mathematical Olympiad, ZIML has a rich history of past participants who have advanced to top tier colleges and prestigious math competitions, including American Math Competitions, MATHCOUNTS, and the International Math Olympaid.

The ZIML Core Online Programs, most available with a free account at ziml.areteem.org, include:

- **Daily Magic Spells:** Provides a problem a day (Monday through Friday) for students to practice, with full solutions available the next day.
- **Weekly Brain Potions:** Provides one problem per week posted in the online discussion forum at ziml.areteem.org. Usually the problem does not have a simple answer, and students can join the discussion to share their thoughts regarding the scenarios described in the problem, explore the math concepts behind the problem, give solutions, and also ask further questions.
- **Monthly Contests:** The ZIML Monthly Contests are held the first weekend of each month during the school year (October through June). Students can compete in one of 5 divisions to test their knowledge and determine their strengths and weaknesses, with winners announced after the competition.
- **Math Competition Practice:** The Practice page contains sample ZIML contests and an archive of AMC-series tests for online practice. The practices simulate the real contest environment with time-limits of the contests automatically controlled by the server.
- **Online Discussion Forum:** The Online Discussion Forum

is open for any comments and questions. Other discussions, such as hard Daily Magic Spells or the Weekly Brain Potions are also posted here.

These programs encourage students to participate consistently, so they can track their progress and improvement each year.

In addition to the online programs, ZIML also hosts onsite Local Tournaments and Workshops in various locations in the United States. Each summer, there are onsite ZIML Competitions at held at Areteem Summer Programs, including the National ZIML Convention, which is a two day convention with one day of workshops and one day of competition.

ZIML Monthly Contests are organized into five divisions ranging from upper elementary school to advanced material based on high school math.

- **Varsity:** This is the top division. It covers material on the level of the last 10 questions on the AMC 12 and AIME level. This division is open to all age levels.
- **Junior Varsity:** This is the second highest competition division. It covers material at the AMC 10/12 level and State and National MathCounts level. This division is open to all age levels.
- **Division H:** This division focuses on material from a standard high school curriculum. It covers topics up to and including pre-calculus. This division will serve as excellent practice for students preparing for the math portions of the SAT or ACT. This division is open to all age levels.
- **Division M:** This division focuses on problem solving using math concepts from a standard middle school math curriculum. It covers material at the level of AMC 8 and School or Chapter MathCounts. This division is open to all students who have not started grade 9.

- **Division E:** This division focuses on advanced problem solving with mathematical concepts from upper elementary school. It covers material at a level comparable to MOEMS Division E. This division is open to all students who have not started grade 6.

This problem book features the Division H Contests. For a detailed list of topics covered for Division H see p.133 in the Appendix.

About Areteem Institute

Areteem Institute is an educational institution that develops and provides in-depth and advanced math and science programs for K-12 (Elementary School, Middle School, and High School) students and teachers. Areteem programs are accredited supplementary programs by the Western Association of Schools and Colleges (WASC). Students may attend the Areteem Institute through these options:

- Live and real-time face-to-face online classes with audio, video, interactive online whiteboard, and text chatting capabilities;
- Self-paced classes by watching the recordings of the live classes;
- Short video courses for trending math, science, technology, engineering, English, and social studies topics;
- Summer Intensive Camps on prestigious university campuses and Winter Boot Camps;
- Practice with selected daily problems for free, and monthly ZIML competitions at ziml.areteem.org.

The Areteem courses are designed and developed by educational experts and industry professionals to bring real world applications into STEM education. The programs are ideal for students who wish to build their mathematical strength in order to excel academically and eventually win in Math Competitions (AMC, AIME, USAMO, IMO, ARML, MathCounts, Math Olympiad, ZIML, and other math leagues and tournaments, etc.), Science Fairs (County Science Fairs, State Science Fairs, national programs like Intel Science and Engineering Fair, etc.) and Science Olympiad, or purely want to enrich their academic lives by taking more challenges and developing outstanding analytical, logical thinking and creative problem solving skills.

Since 2004 Areteem Institute has been teaching with methodology that is highly promoted by the new Common Core State Standards: stressing the conceptual level understanding of the math concepts, problem solving techniques, and solving problems with real world applications. With the guidance from experienced and passionate professors, students are motivated to explore concepts deeper by identifying an interesting problem, researching it, analyzing it, and using a critical thinking approach to come up with multiple solutions.

Thousands of math students who have been trained at Areteem achieved top honors and earned top awards in major national and international math competitions, including Gold Medalists in the International Math Olympiad (IMO), top winners and qualifiers at the USA Math Olympiad (USAMO/JMO), and AIME, top winners at the Zoom International Math League (ZIML), and top winners at the MathCounts National. Many Areteem Alumni have graduated from high school and gone on to enter their dream colleges such as MIT, Cal Tech, Harvard, Stanford, Yale, Princeton, U Penn, Harvey Mudd College, UC Berkeley, UCLA, etc. Those who have graduated from colleges are now playing important roles in their fields of endeavor.

Further information about Areteem Institute, as well as updates and errata of this book, can be found online at http://www. areteem.org.

Acknowledgments

This book contains the Online ZIML Division H Problems from the 2016-17 school year. These problems were created and compiled by the staff of Areteem Institute. These problems were inspired by questions from the Areteem Math Challenge Courses, past questions on the ACT/SAT/GRE, past math competitions, math textbooks, and countless other resources and people encountered by the Areteem Curriculum Department in their life devoted to math. We thank all these sources for growing and nurturing our passion for math.

The Areteem staff, including John Lensmire, David Reynoso, Kevin Wang, and Kelly Ren, are the main contributors who compiled, edited, and reviewed this book. Photographs included on the cover and chapter introduction pages are credit to Kelly Ren and Kevin Wang.

Lastly, thanks to all the students who have participated and continue to participate in the Zoom International Math League. Your dedication to the Daily Magic Spells and Monthly Contests makes all of this possible, and we hope you continue to enjoy ZIML for years to come!

1. ZIML Contests

This part of the book contains the Division H ZIML Contests from the 2016-17 School Year. There were nine monthly competitions, held on the dates found below:

- October 7-8
- November 4-6
- December 2-4
- January 6-8
- February 3-5
- March 3-5
- April 7-9
- May 5-7
- June 2-4

1.1 ZIML October 2016 Division H

Below are the 20 Problems from the Division H ZIML Competition held in October 2016.
The answer key is available on p.146 in the Appendix.
Full solutions to these questions are available starting on p.68.

Problem 1
If $(ax+4)(bx-2) = 6x^2 + cx - 8$ for all values of x, with $a + b = 5$ there are two possible values of c. What is the (positive) difference between the two values of c?

Problem 2
If $a \diamond b \diamond c = \dfrac{a+c}{b}$, what is $3 \diamond 2 \diamond 11$?

Problem 3
Suppose the prices for chicken and turkey per pound are given by

$$c = 1.35 + 0.50x,$$

$$t = 2.85 + 0.35x.$$

where x is the number of weeks after the start of summer. What was the price per pound of turkey when it was equal to the price per pound of chicken?

Problem 4
Calculate $\displaystyle\sum_{k=0}^{5} (-1)^k k^2$.

Problem 5

The sum $(5-2i)+(-1+6i) = A+Bi$ for integers A, B. What is B? Recall $i = \sqrt{-1}$.

Problem 6

Suppose that $ABCD$ is a square, and that CDP is an equilateral triangle, with P outside the square. What is the size of angle PAD in degrees?

Problem 7

Alice and Bob just learned about slope intercept form. Their teacher told them to graph an equation with slope 2 and y-intercept 3. Alice got the equation correct, but Bob mixed up the slope and the y-intercept to get a different line. What is the y-value where the two lines intersect?

Problem 8

Find the global maximum of $f(x) = 61 + 72x - 36x^2$.

Problem 9

If $1 \le x \le 3$, what is the maximum value of $f(x) = 4 - 2x^2$?

Problem 10

Suppose the probability that it will rain on Saturday is $\frac{1}{4}$ and the probability it will rain on Sunday is $\frac{2}{3}$. Suppose further the probability it rains on Sunday is independent of the probability it rains Saturday. The probability it rains both days can be expressed as $\frac{K}{60}$ for an integer K. What is K?

Problem 11

A line in the plane contains the point $(0,2)$ and has slope $-\frac{1}{2}$. For what value of x is $y = 8$.

Problem 12

Suppose $f(x) = 2x - 1$ and $f^{-1}(x)$ is the inverse of $f(x)$. What is $f(5)f^{-1}(-5)$?

Problem 13

Suppose a sphere is inscribed in a cube so that it is tangent to all six faces of the cube. If the sphere has volume 36π, what is the volume of the cube?

Problem 14

Consider the points $O = (0,0,0)$, $A = (2,1,0)$, $B = (2,2,2)$. Set $\angle AOB = \theta$. Then $\cos(\theta) = \frac{\sqrt{M}}{N}$ in simplest radical form. What is $M \div N$?

Problem 15

Suppose $\sin(2\theta) = 0.4$. What is $\sec(\theta)\csc(\theta)$?

Problem 16

If a right circular cone has lateral surface area 12π and a slant height of 3, what is the radius of the base?

Problem 17

The equation $\log_2(x) + \log_4(x) + \log_8(x) = \dfrac{22}{3}$ has one real solution for x. What is this solution?

Problem 18

Four identical red balls and three identical blue balls are arranged in a line. How many different arrangements of the 7 balls are there in total?

Problem 19

Suppose $3x - 2y = 5$. The fraction $\dfrac{8^x}{4^y} = 2^M$ for an integer M. What is M?

Problem 20

A group of pirates went to hunt for treasure. They found a chest of gold coins. They tried to equally divide the coins, but 6 coins were left over. So they picked one pirate among themselves and threw him overboard. Then they tried to divide the coins again, but 5 coins were left over. If the chest held 83 coins, how many pirates were there originally?

1.2 ZIML November 2016 Division H

Below are the 20 Problems from the Division H ZIML Competition held in November 2016.
The answer key is available on p.147 in the Appendix.
Full solutions to these questions are available starting on p.73.

Problem 1
If $\dfrac{x-2}{4} = k$ and $k = -2$, what is the value of x?

Problem 2
Suppose $\dfrac{x}{y} = 52$. What is the value of $\dfrac{x}{4y}$?

Problem 3
Suppose there are 20 girls and 25 boys in a class. They all take a chemistry exam and the class average score is 90. If the average score of the girls is 95, what is the average score of the boys?

Problem 4
8 basketball players form a team. The team must pick 2 players to serve as co-captains for the team. In how many ways can they do so?

Problem 5
What angle (measured in degrees) does the line $y = x\sqrt{3} + 2$ make with the x-axis?

Problem 6

In a circle, suppose $\overarc{AB}$ is an arc with angular measure $60°$ and CD is a diameter such that if rays $\overrightarrow{BA}, \overrightarrow{DC}$ are extended to intersect at a point E, $\angle AEC = 30°$. How many degrees is the angular measure of arc $\overarc{BD}$?

Problem 7

5 yellow and 10 blue balls are in a bag. You randomly pick two different balls (one at a time) from the bag. The probability both balls are blue is $\dfrac{N}{M}$ (the fraction is fully simplified). What is $N + M$?

Problem 8

The complex numbers $3 + 5i$ and $3 - 5i$ are solutions to the quadratic equation $3x^2 - 18x + c = 0$. What is c?

Problem 9

What is the smallest positive integer K such that $\sqrt{150 \cdot K}$ is an integer?

Problem 10

How many ways can you place four identical checkers on the outer edges of an 8×8 checkerboard so that the checkers are in different rows and in different columns?

Problem 11

An isosceles triangle contains a 120 degree angle. If the side opposite the 120 degree angle has length $4\sqrt{3}$, what is the sum of the lengths of the remaining two sides?

Problem 12

How many real roots does the equation

$$(7x - 13)(x^2 + 7x - 13)(x^2 - 7x + 13)$$

have?

Problem 13

Let points $O = (0,0)$ and $Q = (2\sqrt{2},0)$. Point P lies on the parabola $y = 4 - x^2$ so that $OP = PQ$. The area of triangle OPQ is $\sqrt{K}$ for an integer K, what is K?

Problem 14

George has 2 hats, 4 shirts, 1 jacket, 3 pairs of pants, 5 pairs of shorts, and 4 pairs of shoes. George always wears a shirt, shoes, and either shorts or pants (but not both). If he wears pants, he does not wear a jacket or a hat. If he wears shorts, he always wears a hat and possibly wears a jacket. How many different outfits following the above rules can George make?

Problem 15

Consider the equation $|x| - 2 = -|1 - x|$. Let m be the largest solution and n be the smallest solution (with possibly $m = n$ if there is only one solution). What is $m - n$?

Problem 16

In a right triangle, one angle measures $x°$, with $\sin(x) = 0.6$.
What is $\cos(180 - x)$? Give your answer as a decimal.

Problem 17

If $\log_2(x) + \log_2(x^2) = 21$, what is x?

Problem 18

Let triangle ABC be equilateral triangle. Let D be on side $\overline{AB}$ and
E be on side $\overline{AC}$ such that $\overline{DE} \| \overline{BC}$. Let the perimeter of triangle
ADE to be 12 and the perimeter of trapezoid $DECB$ to be 16.
Find the perimeter of triangle ABC.

Problem 19

Calculate

$$\lim_{x \to 2} \frac{x^2 + 2x - 8}{x^2 - x - 2}.$$

Problem 20

How many consecutive zeros are at the end of 30!? Note: for a
positive integer n, the factorial $n!$ is defined as $n! = n \cdot (n-1) \cdot
(n-2) \cdots 2 \cdot 1$.

1.3   ZIML December 2016 Division H

Below are the 20 Problems from the Division H ZIML Competition held in December 2016.

The answer key is available on p.148 in the Appendix.

Full solutions to these questions are available starting on p.79.

Problem 1
What is the constant term of $(x^3 + 2x + 1)(x+2)^2(x^4 - 3x + 2)$?

Problem 2
A right cylinder has a volume of $\dfrac{15\pi}{8}$ cubic feet. If the height of the cylinder is 10 inches, what is the radius of the cylinder in inches?

Problem 3
Find the smallest perfect square that is a multiple of both 75 and 360.

Problem 4
Let $f(x) = 12 - cx^3$, where c is a constant. If $f(2) = 10$, what is the value of $f(-2)$?

Problem 5
Suppose you have a list of 9 elements where each element is an positive integer less than or equal to 10. What is the largest possible difference between the median and mean of this list?

Problem 6
Suppose $a = 3\sqrt{2}$ and $\sqrt{6b} = 4a$. What is the value of b?

Problem 7
Consider two squares attached at their corner vertices as in the diagram below.

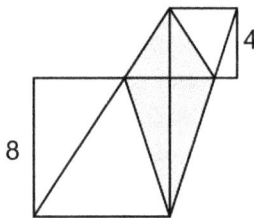

Find the area of the shaded region.

Problem 8
Find the radius of the circle $x^2 + y^2 + 10x - 24y - 87 = 0$.

Problem 9
What is the sum of all possible real solutions to $\sqrt{2-x} - x = 4$?

Problem 10
Suppose you roll a fair 6-sided die 4 times. The probability that all 4 rolls are different is $\dfrac{N}{M}$, where the fraction is written in lowest terms. What is $N + M$?

Problem 11
The equation $4x^2 + 8x + m = 0$ has two distinct real roots for all numbers m if $m < N$ (where N is an integer). What is N?

Problem 12
What is the largest possible angle that can appear in an isosceles triangle containing a $30°$ angle?

Problem 13
Line l contains points $(-4, 0)$ and $(0, 1)$. Line m contains points $(0, -3)$ and $(t, 0)$. If lines l and m are parallel, what is t?

Problem 14
If $0° < x < 90°$ and $4\sin^2(x) = 3 - \sin(x)$, what is the value of $\sin(x)$? Express your answer as a decimal rounded to the nearest hundredth if needed.

Problem 15
A DNA oligomer is a short strand of DNA nucleotides A, C, G, and T. A palindrome is a sequence of letters that is the same forwards and backwards: for example, CCGTATGCC is a palindromic oligomer. How many palindromic oligomers of length 7 are there?

Problem 16

$ABCD$ is a trapezoid and $\overline{AD}$ and $\overline{BC}$ are extended to intersect at E as shown below.

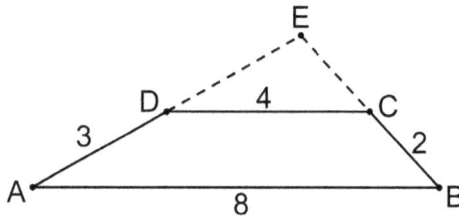

What is the perimeter of triangle CDE?

Problem 17

Let $\lfloor x \rfloor$ denote the greatest integer $\leq x$, and $\lceil x \rceil$ denote the least integer $\geq x$. For example,

$$\lfloor 4.2 \rfloor, \lceil 3.8 \rceil, \lfloor 4 \rfloor, \lceil 4 \rceil$$

are all equal to 4. If $f(x) = 5\lfloor x \rfloor - 3\lceil x \rceil$ and x is not an integer, then $f(x) = A\lfloor x \rfloor + B$ where A, B are integers. What is $A + B$?

Problem 18

A jar contains 99 jelly beans. Some friends pass the jar around clockwise and eat a jelly bean every time they get the jar. Each friend eats at least 2 jellybeans. If the 8th friend is the last one to get a jelly bean, how many friends are there?

Problem 19

Consider numbers n with the property that the factors of n multiply out to n^3. For example, the factors of 12 are $1, 2, 3, 4, 6, 12$ and

$$1 \cdot 2 \cdot 3 \cdot 4 \cdot 6 \cdot 12 = 1728 = 12^3.$$

In fact, 12 is the smallest such number > 1. What is the next smallest number with this property?

Problem 20

When it is defined, $\dfrac{2 + \cos(x)}{1 + 2\sec(x)}$ is equal to $A\cos(x) - B\sin(x)$.
What is $A + B$?

1.4 ZIML January 2017 Division H

Below are the 20 Problems from the Division H ZIML Competition held in January 2017.

The answer key is available on p.149 in the Appendix.

Full solutions to these questions are available starting on p.86.

Problem 1

A batter hits a baseball. The height of the baseball in feet after t seconds is $h = 3 + 64t - 16t^2$. What is the maximum height of the baseball in feet?

Problem 2

A food truck sells burritos for \$8.50 each and tacos for \$3.00 each. In one day they sold a total of 155 burritos and tacos and had a total revenue of \$729.00. How many tacos were sold that day?

Problem 3

A circle with center $(2,2)$ contains the point $(-3,4)$. The equation of this circle can be expressed in the form

$$x^2 + y^2 = Ax + By + C$$

where A, B, C are integers. What is C?

Problem 4

An isosceles trapezoid has area 78 and bases of length 5 and 21. What is the perimeter of the trapezoid?

Problem 5

Suppose an uncle distributes \$1 bills among his nieces and nephews. If he distributed the dollars evenly among the nieces, each niece would get 24 dollars, and if he distributed them evenly among the nephews, each nephew would get 40 dollars. In fact, he distributes the dollars evenly among all his nieces and nephews. How many dollars does each niece or nephew receive?

Problem 6

Suppose there are 10 friends who all live in the same dorm. They receive 4 tickets, to the movies Rocky I, II, III, and IV (so all 4 tickets are different). If each friend gets at most one ticket, how many ways can the tickets be distributed?

Problem 7

Suppose A, B, C are points on a circle such that the angular measures of arc $\overarc{AB}$ (not containing C) and arc $\overarc{CA}$ (not containing B) are in ratio $5 : 9$. Suppose further that $\angle ABC = 90°$. Find the measure of $\angle BAC$ in degrees.

Problem 8

How many perfect squares are factors of 2592?

Problem 9

Consider integers x with $-5 \leq x \leq 5$. For how many of these integers is $3x - 5 \geq 4x - 3$?

Problem 10

Let $ABCD$ be a parallelogram, and E,H,F,G be points on sides $\overline{AB},\overline{BC},\overline{CD},\overline{DA}$ respectively, and $\overline{EF}\|\overline{BC}$ and $\overline{GH}\|\overline{AB}$. Let P be the intersection of $\overline{EF}$ and $\overline{GH}$. If $[GPFD]=10,[PHCF]=8,[EBHP]=16$, find $[ABCD]$. (Here $[-]$ is used to denote area, so for example $[GPFD]=10$ means the area of rectangle $GPFD$ is 10.)

Problem 11

a and b are such that (for $x>1$)

$$\frac{x^{a^2}}{x^{b^2}}=x^{28} \text{ and } \frac{x^b}{x^a}=x^7.$$

What is $a+b$?

Problem 12

What is the (positive) difference between the internal and external angles of a regular pentagon? Give your answers in degrees.

Problem 13

$\log_{\sqrt{3}}(x)+2\log_3(x)-\log_9(x)$ can be written as $\log_3(x^r)$ where r is a reduced fraction of the form $\dfrac{p}{q}$. What is $p+q$?

Problem 14

Consider a circle divided into 5 regions as in the diagram below, where the size of each region is proportional to its number.

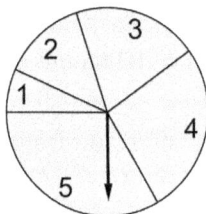

The spinner is spun around the center of the circle and randomly stops in one of the regions. The probability the spinner stops in the region labeled '4' is $\dfrac{N}{M}$ where the fraction is in lowest terms. What is $M - N$?

Problem 15

In $\triangle ABC$, suppose $AB = 15$, $BC = 13$ and $\cos B = \dfrac{33}{65}$. $\sin C$ can be written as $\dfrac{P}{Q}$ with $\gcd(P, Q) = 1$. What is $P + Q$?

Problem 16

Four identical balls (spheres), each of radius 1 inch, are glued to the ground so that their centers form the vertices of a square with side length 2 inches. Suppose you rest a fifth identical ball on the four balls (so the fifth ball is a sphere externally tangent to the other spheres). The fifth ball rests $\sqrt{L}$ inches off the ground. What is L?

Problem 17

Let $f(z) = 6z^5 + 5z^4 + 4z^3 + 3z^2 + 2z + 1$. Then $f(i) = A + Bi$ for integers A, B. What is $A^2 + B^2$? Note: Here $i = \sqrt{-1}$.

Problem 18

Suppose you have a list of 5 integers, each between 1 and 10 (inclusive). It is possible that the same number appears more than once. How many different averages for the list are possible?

Problem 19

From the set of natural numbers $2, 3, 4, \ldots, 999$, delete 9 subsets as follows. First delete all even numbers except 2, then all multiples of 3, except 3, then all multiples of 5, except 5, and so on for the nine primes $2, 3, 5, 7, 11, 13, 17, 19, 23$. After all these steps, there are only three composite numbers left. Find the sum of these composite numbers.

Problem 20

Solve $2 \log_2(x) \log_4(x) + 2 \log_2(x) - 3 = 0$. What is the product of all the solutions? Round your answer to the nearest hundredth if necessary.

1.5   ZIML February 2017 Division H

Below are the 20 Problems from the Division H ZIML Competition held in February 2017.
The answer key is available on p.150 in the Appendix.
Full solutions to these questions are available starting on p.93.

Problem 1

Alice and Bob just learned about slope intercept form. Their teacher told them to graph an equation with slope 3 and y-intercept -2. Alice got the equation correct, but Bob mixed up the slope and the y-intercept to get a different line. Alice's line and Bob's line intersect when $y = K$ for an integer K. What is K?

Problem 2

For $|x| \leq 5$, what is the maximum value of $f(x) = x^2 - 4x + 1$?

Problem 3

In the diagram, the area of region A equals the area of region B plus $50\pi - 100$. Find the height of the triangle.

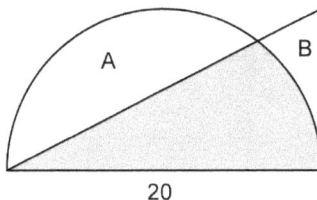

Problem 4

The SAT is a renowned high school standardized test administered for college admissions purposes. The scores are in multiples of 10 (maximum 800 per section) and the total score is determined by adding the critical reading subscore and the math subscore. Of the 50 students that scored above 600 on either the critical reading or math section of the SAT, 35 students scored above 600 on the math section and 25 students scored above 600 on the critical reading section. What is the maximum average total score of all 50 students?

Problem 5

Recall that $\lfloor x \rfloor$ is the greatest integer $\leq x$. What is the sum of all solutions to $\lfloor x \rfloor^2 + \lfloor x \rfloor + 1 = 5x - 2$?

Problem 6

Suppose in triangle $\triangle ABC$, $b = 4$ and $\angle B = 45°, \angle C = 75°$. Then the length of side a is $R\sqrt{S}$ written in simplest radical form. What is $R + S$?

Problem 7

Consider the line $y = 2x + 5$. The closest this line gets to the origin $(0,0)$ is $\sqrt{D}$. What is D?

Problem 8

Suppose you have a box (rectangular prism) that is 2 feet long, 1 foot wide, and has a height of 6 inches. You want to double the volume of the box by changing one of the dimensions of the box. Consider the surface area of the possible new boxes. What is the sum of all the different possible surface areas? Give your answer in square feet.

Problem 9

What is the sum of all real solutions to $\sqrt{x+3} - \sqrt{3x-2} = -1$?
Note: If there are no real solutions, input 0.

Problem 10

Suppose five people are to be seated in a row of 9 chairs. How many possible seating arrangements can be made if there must be a seat in between each person?

Problem 11

Suppose $\sin(\theta) = \dfrac{5}{13}$ where $\dfrac{\pi}{2} < \theta < \pi$. Then $\tan(2\theta) = \dfrac{N}{M}$ with $\gcd(N,M) = 1$ and $M > 0$. What is $M - N$?

Problem 12

A prison houses $1,000$ inmates in $1,000$ prison cells. It also has $1,000$ guards. One day the guards decide to use a number theory problem to free some prisoners. The first guard goes through the prison and unlocks every cell. The second guard goes through the prison and locks every second cell. The third guard goes through the prison and changes the status of every third cell. That is, if it's locked, he unlocks it; and if it's unlocked, he locks it. This process continues (that is the nth prison guard changes the status of every nth cell) until the $1,000$th guard has passed through the prison, after which every prisoner whose cell is left unlocked is free to leave. How many prisoners will be set free?

Problem 13

Solve the equation

$$\log_{\sqrt{3}} x + \log_3 x + \log_9 x = -\frac{21}{2}.$$

The smallest positive solution can be written as $\frac{P}{Q}$ with $\gcd(P,Q) = 1$. What is $P + Q$?

Problem 14

Two perpendicular chords intersect and divide each other in a ratio of $1 : 2$. If the two chords each have length 12, the area of the circle is $\pi \cdot L$ where L is an integer. What is L?

Problem 15

Suppose $f(x) = 3x + 2$ and $f^{-1}(x)$ is the inverse of $f(x)$. What is $f(7)f^{-1}(-7)$?

Problem 16

Suppose a right triangle has integer length sides in ratio $3 : 4 : 5$. Suppose further that the altitude to the hypotenuse is also an integer. What is the smallest possible area for such a triangle?

Problem 17

Write $(2+i)^4$ in the form $S + Ti$. What is $S^2 + T^2$?

Problem 18

Suppose you roll a fair six-sided die. You then flip a coin the number of times shown on the die. The probability you get exactly 5 heads (we do not care how many tails) can be written as a reduced fraction $\dfrac{N}{M}$. What is M?

Problem 19

In triangle ABC, $AB = c$, $BC = a$, and $CA = b$. Suppose that $(a+b+c)(a+b-c) = 3ab$. Determine the measure of $\angle C$ in degrees.

Problem 20

What is the ones digit of $1^2 + 2^2 + 3^2 + \cdots + 99^2$?

1.6 ZIML March 2017 Division H

Below are the 20 Problems from the Division H ZIML Competition held in March 2017.
The answer key is available on p.151 in the Appendix.
Full solutions to these questions are available starting on p.101.

Problem 1
If $3s + 5t = 10$ and $2s - t = 7$, what is the value of $\frac{1}{2}s + 3t$?

Problem 2
The area of parallelogram $ABCD$ shown below is 24. If $AB = 6$ and the drawn height of the parallelogram divides the base into two equal lengths, what is the perimeter of $ABCD$?

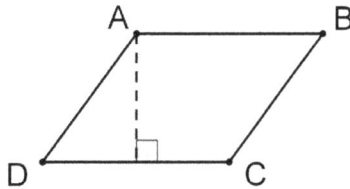

Problem 3
Let $f(x) = \frac{1}{3}x - 3$. Find the sum of all possible values of x satisfying $f(x)f^{-1}(x) = 0$.

Problem 4

Two fair dice are rolled simultaneously. Calculate the product of the two numbers rolled. The probability that this product is at least 24 can be expressed as a fraction $\dfrac{p}{q}$ in lowest terms. What is p+q?

Problem 5

Draw the shortest line segment possible connecting the point $(3,4)$ to the circle with equation $x^2 + y^2 = 9$. How long is this line segment?

Problem 6

For what value of x does

$$2^x + 4^x + 8^x = 584?$$

Problem 7

Consider two-digit numbers whose square ends in the same two-digit number. Find the sum of all possible such numbers. Note: 01 is not a two-digit number.

Problem 8

In the diagram below, the triangle is a right triangle, and both circular arcs have centers that lie on the edges of the triangle.)

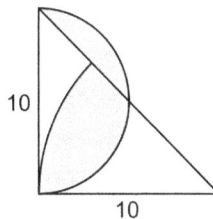

The total area of the shaded regions can be written as $R \times \pi + S$. What is $R + S$?

Problem 9

Suppose you have a list of 5 integers chosen from $1, 2, 3, \ldots, 10$. If it is possible for the same number to appear in the list more than once, how many different means (averages) are possible?

Problem 10

Sally and Megan are running on a park. Starting at the same point, they both start running at the same time at the intersection of two perpendicular paths. Sally runs north and Megan runs east twice as fast as Sally. After 10 minutes they are 2.5 miles apart of each other. Megan's speed in miles per hour is $A\sqrt{B}$, where A and B are integers and the square root cannot be simplified. What is $A + B$?

Problem 11

Solve $|x - |2x + 1|| = 3$. What is the sum of all possible solutions? Round your answer to the nearest hundredth if necessary.

Problem 12

In the following triangle, find the value of θ to the nearest degree.

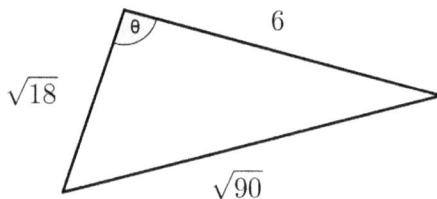

Problem 13

Suppose p and q are integers and the quadratic equation $x^2 + px + q = 0$ has two complex roots $2 + ai$ and $b + i$. What is q?

Problem 14

Suppose A, B, C are integers ≥ 2 with (i) $\gcd(A, B) = 12$, (ii) $\text{lcm}(A, B) = 396$, and (iii) $\gcd(B, C) = 33$. Calculate $\gcd(11A, B)$.

Problem 15

Suppose there is an isosceles triangle containing a 120 degree angle. If the side opposite the 120 degree angle has length $4\sqrt{3}$, what is the sum of the lengths of the other two sides?

Problem 16
Find the sum of all possible values of x satisfying $\log_x(4x-3)=2$.

Problem 17
Evaluate the following: $\dfrac{(2017^2-2023)(2017^2+4031)\times 2018}{2014\cdot 2016\cdot 2018\cdot 2020}$.

Problem 18
Of the 12 members of a high school drama club, 7 are seniors. The club plans to establish an 8 member committee to interview potential club members. If exactly 5 members of the committee must be seniors, how many committees are possible?

Problem 19
Suppose you have an ice cream cone with radius 2 inches and height 4 inches. The cone starts full of ice cream (but there is not ice cream outside the cone). After you've eaten some ice cream and some of the cone you are left with a cone with a radius and a height of 2 inches. What percentage of the ice cream have you eaten? For example, if you have eaten 10% of the ice cream, input 10 as your answer.

Problem 20
Consider solutions to $\sin(2x)=\tan(x)$, where x is measured in degrees. What is the smallest solution with $x>180°$?

1.7 ZIML April 2017 Division H

Below are the 20 Problems from the Division H ZIML Competition held in April 2017.
The answer key is available on p.152 in the Appendix.
Full solutions to these questions are available starting on p.109.

Problem 1
Paul bought a basil plant that started out 2 inches tall. Greg planted a basil plant seed (so started 0 inches tall). The height of Paul's basil plant in inches is modeled by the function $f(x) = 2 + 3x$ where x is the number of weeks. The height of Greg's plant is modeled by the function $g(x) = x^2 + x$. After how many weeks will Paul's basil and Greg's basil have the same height? Express your answer as a decimal rounded to nearest tenth.

Problem 2
Suppose that $ABCD$ is a square, and that CDP is an equilateral triangle, with P inside the square. What is the size of $\angle PAD$, measured in degrees?

Problem 3
Farmer Billy has a farm where he raises chickens and rabbits. One Sunday Billy counted 500 legs on the farm (2 legs per chicken and 4 legs per rabbit.) On Monday he traded all his rabbits for chickens (1 rabbit for 1 chicken). On Tuesday, he traded all his original chickens (not the new ones) so that he received 3 rabbits for every 2 chickens. On Wednesday Billy remarked that there were still 500 legs on the farm. How many chickens did Billy start the week with?

Problem 4

As shown in th diagram below, in quadrilateral $ABCD$, $\overline{AE}$ and $\overline{AF}$ are angle bisectors of $\angle BAC$ and $\angle DAC$ respectively.

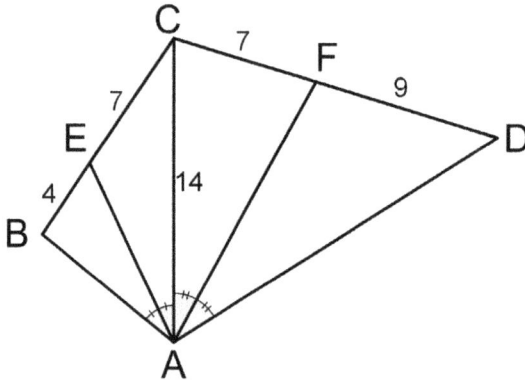

Given the lengths of the segments as marked in the diagram, find the perimeter of $ABCD$.

Problem 5

Jamie started off the semester with the following six grades on her quizzes:

$$70, 75, 75, 85, 85, 90.$$

What is the minimum number of quizzes Jamie must take to increase her average to at least 90? Assume that the maximum score for any quiz is 100.

Problem 6

What is the units/ones digit of $14^{20} + 16^{30}$?

Problem 7

A quadratic $y = Ax^2 + Bx + C$ has a minimum value of 5. Further suppose the points $(0,8)$ and $(-6,8)$ are on the quadratic. What is $A + B + C$? Round your answer to the nearest hundredth if necessary.

Problem 8

The region bounded by $y = x$, $y = 6 - x$, and $y = \frac{1}{2}x$ is a triangle. What is the area of this triangle?

Problem 9

The function $f(x) = |x^2 - 4|$ does not have an inverse, but can be broken into pieces so that each piece has an inverse. What is the fewest number of pieces you need to break $f(x)$ into so that each piece has an inverse?

Problem 10

Suppose in $\triangle ABC$ (with sides a, b, c opposite from $\angle A, \angle B, \angle C$) we know $a = 5$, $\angle A = 30°$, and $\angle B = 45°$. What is side b? If necessary round your answer to the nearest tenth.

Problem 11

How many numbers from $1, 2, 3, 4, \ldots, 200$ have an odd number of factors?

Problem 12
A square pyramid with base length 6 and height 4. If V is the volume and S is the surface area of the pyramid, calculate $V - S$.

Problem 13
How many ways are there to rearrange all the letters from the word *BANANAS*?

Problem 14
Suppose $\overset{\frown}{AB}$ is an arc with angular size $50°$ and CD is a diameter such that if rays $\overrightarrow{BA}, \overrightarrow{DC}$ are extended to intersect at a point E. If $\angle AEC = 20$, find the angular size of arc $\overset{\frown}{BD}$ in degrees. (Recall that a ray $\overrightarrow{BA}$ starts at point B and contains the point A.)

Problem 15
Consider the points $A = (0,0,0)$, $B = (3,4,0)$, and $C = (2,2,1)$. Then $\cos \angle BAC = \dfrac{P}{Q}$ for P, Q positive integers and $\gcd(P,Q) = 1$. What is $P + Q$?

Problem 16
Let $x, y, z > 1$ be such that $\log_{xy}(z) = 2$ and $\log_z(x/y) = 1$. What is $\log_y(z)$?

Problem 17

You are given that $x = i$ and $x = -1 + i$ are roots of $p(x) = x^6 - 3x^4 - 8x^3 - 8x^2 - 8x - 4$. How many real roots does $p(x)$ have?

Problem 18

How many zeros are at the end of $200!$? Recall $200! = 200 \times 199 \times 198 \times \cdots \times 2 \times 1$.

Problem 19

In the equation $x^2 - 101x + k = 0$, one of the roots plus 4 equals 20 times the other root. What is k?

Problem 20

At a track meet, Jared needs to jump longer that 3 meters to qualify for the next meet in long jump. He has two attempts to do so (and only needs one jump longer than 3 meters to advance). Assume that for each attempt, Jared randomly jumps a distance in the interval $[1,4]$ (all real numbers from 1 to 4 (inclusive) are possible). The the two jumps do not affect each other, the probability that Jared qualifies for the next meet is $\dfrac{P}{Q}$ for positive integers $P, Q > 0$ with $\gcd(P, Q) = 1$. What is $P + Q$?

1.8 ZIML May 2017 Division H

Below are the 20 Problems from the Division H ZIML Competition held in May 2017.

The answer key is available on p.153 in the Appendix.

Full solutions to these questions are available starting on p.117.

Problem 1
A rectangle has perimeter 24 and area 35. Find the difference between the larger and smaller side length.

Problem 2
A parabola has equation $y = x^2 + bx + c$ and the line $y = 5$ intersects the parabola at $x = -1, 3$. What is the minimum value of this parabola?

Problem 3
Find the units digit of 3^{2017}.

Problem 4
Suppose $2 + 3i$ is a solution to the quadratic equation $4x^2 - 16x + C = 0$ for an integer C. What is C?

Problem 5

In the diagram, $\triangle ABC, \triangle DEF$ are two congruent isosceles right triangles. Given that $AB = 9, EC = 3$, find the area of the shaded region.

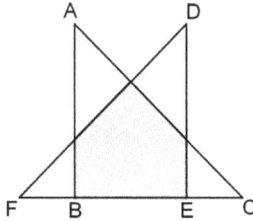

Problem 6

6 boys and 4 girls run a race. Suppose the boys finish in alphabetical order based on their names (and each boy has a different name). If there are no ties and everyone finishes the race, how many different outcomes of the race are there?

Problem 7

What is the coefficient of x^8 in $(x^3 + 2x + 1)(x + 2)^2(x^4 - 3x + 2)$.

Problem 8

Suppose you have a circle $(x - 1)^2 + (y - 1)^2 = 4$ with center C and a line $x + y = 4$. This line intersects the circle at two points, call them A and B. Use points A, B, and C to form a triangle $\triangle ABC$. What is the area of this triangle? Round your answer to the nearest tenth if necessary.

Problem 9

A dog is leashed at the top corner of a building whose base is an equilateral triangle with side length 3 m, as shown in the diagram.

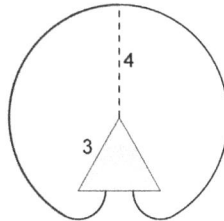

The length of the rope is 4 m. The total area of the region (in m^2) that the dog can reach can be expressed as $K \times \pi$ for a positive K. What is K, rounded to the nearest integer if necessary?

Problem 10

Suppose you have a list of 6 numbers each chosen from $1, 2, 3, \ldots, 10$. If it is possible the same number to appear in the list more than once, how many different medians are possible?

Problem 11

For how many integer values of m does $x^2 - 4x - m = 0$ have no real solutions while $x^2 - 18x + m^2 = 0$ has at least one real solution?

Problem 12

Points A, B, C, D lie in the plane. Suppose C lies between A, B on $\overline{AB}$ while D is not on line $\overleftrightarrow{AB}$. If $CD = BC = 3, AC = 6$, and $BD = 4$, what is AD?

Problem 13

The equation $\log_3(x) + \log_9(x) + \log_{27}(x) = \dfrac{22}{3}$ has one integer solution. Find this solution.

Problem 14

There are 9 divisors for number A and 10 divisors for number B. The least common multiple of A and B is 2800. What is the larger of the two numbers?

Problem 15

Consider lines $\ell_1 : y = x - 2, \ell_2 : y = x + 2, \ell_3 : y = 6 - x$. Find the square of the radius of largest circle such that (i) The center of the circle is on line ℓ_3 and (ii) The circle does not go outside lines ℓ_1 and ℓ_2.

Problem 16

Consider solutions to

$$2 \sin (2x - 90°) = \sqrt{2}$$

where x is measured in degrees and $0° \leq x < 360°$. What is the sum of all such solutions?

Problem 17

Consider solutions to $x^4 - 3x^3 - x^2 + 9x - 6 = 0$. Two solutions are in the form $A \pm \sqrt{B}$ for integers A, B. What is $A + B$?

Problem 18

Kary, Larry, and Mary describe a cone in three-dimensions.
Kary says that the center of the base is the point $(1, 1, 1)$.
Larry says that the height of the cone is 10.
Mary says that the circular edge of the base contains the point $(-2, 3, -5)$.
The volume of this cone is $C \times \pi$. What is C, rounded to the nearest hundredth if necessary?

Problem 19

Suppose you randomly pick a real number (call it x) from the closed interval $[-5, 5]$. Find the probability that $|x - 1| > 2$. Express your answer as $P\%$ where, P is rounded to the nearest integer.

Problem 20

Recall $\lfloor x \rfloor$ is the greatest integer $\leq x$ and $\{x\} = x - \lfloor x \rfloor$. The equation $x + \{x\} = 2 \lfloor x \rfloor$ has one solution where $x \neq 0$. What is this solution? Round your answer to the nearest hundredth if necessary.

1.9 ZIML June 2017 Division H

Below are the 20 Problems from the Division H ZIML Competition held in June 2017.
The answer key is available on p.154 in the Appendix.
Full solutions to these questions are available starting on p.124.

Problem 1
The mean width of 12 iPads is 5.2 inches. The mean width of 8 Kindles is 4.7 inches. What is the mean width of the 12 iPads and 8 Kindles in inches?

Problem 2
How many positive integers between 1 and 1000 (inclusive) are multiples of either 13 or 17 or both?

Problem 3
Consider the diagram below:

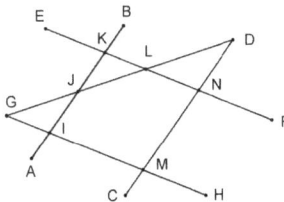

Suppose we know that $\overleftrightarrow{AB}$ and $\overleftrightarrow{CD}$ are parallel, $\angle DMH = 70°$, $\angle ELD = 135°$, and $\angle AJG = 30°$. What is $\angle LND$ in degrees?

Problem 4

Bill, Claire, and Drew went on a road trip over the weekend. In total they drove 485 miles over 10 hours. Bill drove at an average rate of 40 mph, Claire an average of 50 mph, and Drew an average of 60 mph. If Bill and Drew drove the same distance, how many hours did Claire drive? Round your answer to the nearest hundredth if necessary.

Problem 5

In the diagram below, there are 21 grid points arranged in equilateral triangles, equally spaced.

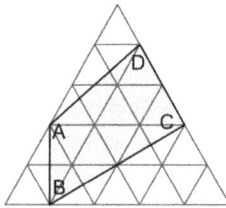

The area of each small equilateral triangle formed by 3 adjacent grid points is 1. Find the area of quadrilateral $ABCD$.

Problem 6

Find the smallest m so that $f(x) = x^2 - 6x + 12$ for $x \geq m$ has an inverse. What is m?

Problem 7

How many times does the parabola $y = 2x^2 - 8x + 10$ intersect the circle $x^2 + y^2 = 4x$?

Problem 8

Let $ABCD$ be a parallelogram, and E, H, F, G be points on sides $\overline{AB}, \overline{BC}, \overline{CD}, \overline{DA}$ respectively, with $\overline{EF} \| \overline{BC}$ and $\overline{GH} \| \overline{AB}$. Let P be the intersection of $\overline{EF}$ and $\overline{GH}$. If $[GPFD] = 10, [PHCF] = 8, [EBHP] = 16$, find $[ABCD]$. (We use $[WXYZ]$ to represent the area of quadrilateral $WXYZ$, so, for example, the area of quadrilateral $GPFD$ is 10.)

Problem 9

Two friends decide to trade some of their books. The first friend has a total of 6 books and the second has 8 books. How many ways can they trade 3 books from the first friend for 3 books from the second friend? We do not care about the order of the books traded, just which books transfer between friends.

Problem 10

In the $2 \times 2 \times 2$ cubic figure below, consider the shortest path from A to B along the surface of the cube.

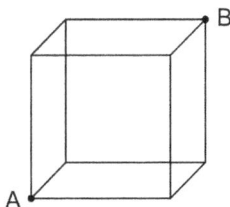

The length of this shortest path can be written in simplest radical form $P\sqrt{Q}$. What is $P + Q$?

Problem 11

Note $x = \pm 1$ are solutions to $\dfrac{1}{x^2} - \dfrac{1}{x^2 + 3} = \dfrac{3}{4}$. Two imaginary solutions also exist, with form $\pm K \times i$ for an integer K. What is K?

Problem 12

Suppose in triangle $\triangle ABC$ we have $\angle A = 45°$, $AB = 3$, and $AC = 2\sqrt{2}$. What is $\tan(C)$? Round your answer to the nearest tenth if necessary.

Problem 13

Suppose you pick a number from $\{1,2,3,4,5,6\}$ with probabilities:

$$P(1) = P(3) = 0.2, P(2) = P(4) = 0.1, P(5) = 0.15.$$

(So, for example, the probability of picking 1 is 0.2.) The probability you pick an even number can be written as $K\%$ where K is an integer. What is K?

Problem 14

Let $ABCD$ be a square with edge length of 2 cm. We draw a circle of center A and radius 2 cm and a circle of center C and radius 2 cm. Consider the region that is inside both circles (and therefore also inside the square). What is the area of this region in square cm? Round your answer to the nearest hundredth.

Problem 15

What is the sum of all solutions to $\log_x(2) + \log_2(x) = \dfrac{5}{2}$? If necessary, round your answer to the nearest tenth.

Problem 16

There is one triple (x, y, z) with x, y, z different positive prime numbers satisfying the equation $x(x+y)(y+z) = 140$. For this triple, what is $x + y + z$?

Problem 17

Find the area of a triangle with vertices $A = (0,0,0)$, $B = (3,0,0)$ and $C = (3, 2\sqrt{2}, 2\sqrt{2})$. Round your answer to the nearest tenth if necessary.

Problem 18

A natural number is a multiple of 72, and has a total of 15 factors. Find the largest such number.

Problem 19

Consider the equation $\sin(2x) = \tan(x)$ where x is measured in degrees. How many solutions does this equation have with $0° \le x < 360°$?

Problem 20

Write $\left(\dfrac{\sqrt{2}}{2}(-1+i) \right)^{100}$ in the form $A + Bi$. What is $B - A$?

2. ZIML Solutions

This part of the book contains the official solutions to the problems from the nine Division H ZIML Contests from the 2016-17 School Year.

Students are encouraged to discuss and share their own methods to the problems using the Discussion Forum on ziml.areteem.org.

2.1 ZIML October 2016 Division H

Below are the solutions from the Division H ZIML Competition held in October 2016.

The problems from the contest are available on p.15.

Problem 1 Solution

Expanding we have $abx^2 + (-2a+4b)x - 8 = 6x^2 + cx - 8$ so $a \times b = 6$. Since $a + b = 5$ as well, we have $a = 2, b = 3$ or $a = 3, b = 2$. Since $c = -2a + 4b$ we can plug in these two possibilities to get $c = 8$ or $c = 2$. Thus our final answer is $8 - 2 = 6$.

Answer: 6

Problem 2 Solution

Using the formula we have $3\diamond2\diamond11 = \dfrac{3+11}{2} = 7$.

Answer: 7

Problem 3 Solution

If the prices are equal, then $c = t$ or $1.35 + 0.5x = 2.85 + 0.35x$. Combining like terms gives us $0.15x = 1.50$ so $x = 10$. Plugging back in we find the price per pound of chicken is $2.85 + 0.35 \cdot 10 = 6.35$.

Answer: 6.35

Problem 4 Solution

We can directly calculate

$$\sum_{k=0}^{5}(-1)^k k^2 = (-1)^0 \cdot 0^2 + (-1)^1 \cdot 1^2 + \cdots + (-1)^5 \cdot 5^2$$

$$= -1 + 4 - 9 + 16 - 25 = -15$$

to get the correct answer.

Answer: -15

Problem 5 Solution
We have $(5 - 2i) + (-1 + 6i) = (5 - 1) + (-2 + 6)i = 4 + 4i$. Hence $B = 4$.

Answer: 4

Problem 6 Solution
We have $\angle ADP = \angle ADC + \angle CDP = 90 + 60 = 150°$. Since $\triangle ADP$ is isosceles $\angle PAD = (180 - 150)/2 = 15°$.

Answer: 15

Problem 7 Solution
Alice got the correct line of $y = 2x + 3$ (in slope-intercept form). Bob mixed up so he got $y = 3x + 2$. Setting these the y values equal gives $2x + 3 = 3x + 2$ so $x = 1$ after combining like terms. Plugging back in gives $y = 2(1) + 3 = 5$, so they intersect at $(1, 5)$.

Answer: 5

Problem 8 Solution
The vertex is $-72/(2 \cdot (-36)) = 1$ and the x^2 coefficient is negative, hence $f(1) = 97$ is a maximum.

Answer: 97

Problem 9 Solution
$f(x)$ is a parabola opening downwards with vertex $(0, 4)$. Hence $f(x)$ is decreasing for all $x > 0$. Thus the maximum in the interval

$[1,3]$ is at $x = 1$, which is $f(1) = 4 - 2 \cdot 1^2 = 2$.

Answer: 2

Problem 10 Solution

Since the two days are independent, we can multiply the probabilities: $\dfrac{1}{4} \cdot \dfrac{2}{3} = \dfrac{1}{6} = \dfrac{10}{60}$, so $K = 10$.

Answer: 10

Problem 11 Solution

The line has slope $-1/2$ and y-intercept 2, so has equation $y = \dfrac{-x}{2} + 2$. Setting $y = 8$ we have $8 = \dfrac{-x}{2} + 2$ so solving for x we get $x = 6 \cdot (-2) = -12$.

Answer: -12

Problem 12 Solution

Set $y = 2x - 1$ and swap x, y to find the inverse: $x = 2y - 1$ so $y = \dfrac{x+1}{2}$. Thus $f^{-1}(x) = \dfrac{x+1}{2}$. Since $f(5) = 9$ and $f^{-1}(-5) = -2$, their product is -18.

Answer: -18

Problem 13 Solution

The sphere has volume 36π. Since the volume of a sphere is given by $\dfrac{4}{3}\pi r^3$ we can solve for r: $\dfrac{4}{3}\pi r^3 = 36\pi$ so $r^3 = 27$ and hence $3 = r$. Hence the diameter of the sphere is 6, which is also the side length of the cube. Hence the cube has volume $6^3 = 216$.

Answer: 216

Problem 14 Solution

Using the distance formula $OA = \sqrt{2^1 + 1^2 + 0^1} = \sqrt{5}$, $OB = \sqrt{2^2 + 2^2 + 2^2} = 2\sqrt{3}$, $AB = \sqrt{0^2 + 1^2 + 2^2} = \sqrt{5}$. Since we know all three sides of $\triangle AOB$ we can use the Law of Cosines, which says $AB^2 = OA^2 + OB^2 - 2 \cdot OA \cdot OB \cdot \cos(\theta)$. Thus $5 = 5 + 12 - 2 \cdot \sqrt{5} \cdot 2\sqrt{3} \cdot \cos(\theta)$ so $-12 = -4\sqrt{15}\cos(\theta)$ so $\cos(\theta) = \dfrac{3}{\sqrt{15}} = \dfrac{\sqrt{15}}{5}$. Hence $M = 15, N = 5$ so $M \div N = 3$.

Answer: 3

Problem 15 Solution

Recall using the double angle formula

$$0.4 = \sin(2\theta) = 2\sin(\theta)\cos(\theta).$$

Rewriting the second equation in terms of sine and cosine we have

$$\sec(\theta)\csc(\theta) = \frac{1}{\cos(\theta)\sin(\theta)} = \frac{2}{2\sin(\theta)\cos(\theta)} = \frac{2}{0.4} = 5.$$

Answer: 5

Problem 16 Solution

Recall a right circular cone has lateral surface area $\pi \cdot r \cdot l$ where r is the radius and l is the slant height (recall $l = \sqrt{r^2 + h^2}$ where h is the height of the cone). Hence we have $12\pi = \pi \cdot r \cdot 3$ so $r = 4$.

Answer: 4

Problem 17 Solution

We have (using change of base)

$$\log_2(x) + \log_4(x) + \log_8(x) = \log_2(x) + \frac{\log_2(x)}{\log_2(4)} + \frac{\log_2(x)}{\log_2(8)}$$

$$= \log_2(x) + \frac{\log_2(x)}{2} + \frac{\log_2(x)}{3} = \frac{11\log_2(x)}{6}.$$

Hence we need $\log_2(x)\dfrac{22}{3} \cdot \dfrac{6}{11} = 4$ so $x = 2^4 = 16$.

Answer: 16

Problem 18 Solution

Consider 7 total positions for the balls. There are 4 identical, so we must choose (without order) 4 positions for them. This can be done in $\dbinom{7}{4} = {_7}C_4 = \dfrac{7!}{3! \cdot 4!} = 35$ ways. The 3 identical blue balls must take up the remaining 3 positions, so there are 35 arrangements in total.

Answer: 35

Problem 19 Solution

Solving $3x - 2y = 5$ for y we get $y = \dfrac{1}{2}(3x - 5)$. Then note that $8 = 2^3$ and $4 = 2^2$ so we can use the rules for exponents to get

$$\frac{8^x}{4^y} = \frac{2^{3x}}{2^{2y}} = \frac{2^{3x}}{2^{3x-5}} = 2^{3x - (3x - 5)} = 2^5.$$

Hence $M = 5$ is our final answer.

Answer: 5

Problem 20 Solution

$83 - 6 = 77$, so the number of pirates at the beginning was a factor of 77. $83 - 5 = 78$, so after shooting one pirate, the remaining number of pirates is a factor of 78. Looking at the factors of the numbers 77: $1, 7, 11, 77$ and 78: $1, 2, 3, 6, 13, 26, 39, 78$, we conclude that there were 7 pirates originally, and 6 pirates now.

Answer: 7

2.2 ZIML November 2016 Division H

Below are the solutions from the Division H ZIML Competition held in November 2016.

The problems from the contest are available on p.21.

Problem 1 Solution

Cross multiplying we have $x - 2 = 4 \cdot -2 = -8$. Hence $x = -8 + 2 = -6$.

Answer: -6

Problem 2 Solution

We have that

$$\frac{x}{4y} = \frac{x}{y} \div 4 = 52 \div 4 = 13.$$

Answer: 13

Problem 3 Solution

Since the average of all 45 students is 90, all the scores combined sum to

$$45 \times 90 = 4050.$$

The average of the girls is 95 so the sum of all their scores is

$$20 \times 95 = 1900.$$

Hence the sum of all the scores of the 25 boys is $4050 - 1900 = 2150$. Therefore the average of the boys is

$$\frac{2150}{25} = 86.$$

Answer: 86

Problem 4 Solution

Since the order of the co-captains does not matter, there are

$$\binom{8}{2} = {}_8C_2 = \frac{8 \cdot 7}{2} = 28$$

possibilities.

Answer: 28

Problem 5 Solution

The slope is $\sqrt{3}$. Since the slope is $\Delta y / \Delta x$ we want an angle θ such that $\tan(\theta) = \sqrt{3}$. Therefore $\theta = 60°$.

Answer: 60

Problem 6 Solution

Let the angular measure of $\overset{\frown}{BD} = x$. Then the size of $\overset{\frown}{AC} = 180 - 60 - x = 120 - x$. Then $\angle AEC = 30 = \dfrac{x - (120 - x)}{2}$ so solving for x gives $x = 90°$.

Answer: 90

Problem 7 Solution

In total there are $15 \times 14 = 210$ outcomes. There are $10 \times 9 = 90$ outcomes where both are blue. Hence the probability both are blue is

$$\frac{90}{210} = \frac{9}{21} = \frac{3}{7}$$

so the answer is $3 + 7 = 10$.

Answer: 10

Problem 8 Solution

Using the quadratic equation we see that the product of $3 + 5i$ and $3 - 5i$ must be equal to $\frac{c}{3}$. Hence

$$c = 3 \cdot (3 + 5i)(3 - 5i) = 3 \cdot (9 + 25) = 3 \cdot 34 = 102.$$

Answer: 102

Problem 9 Solution

Note that $150 = 2 \cdot 3 \cdot 5^2$. Hence if $K = 2 \cdot 3 = 6$ we have

$$\sqrt{150 \cdot K} = \sqrt{2^2 \cdot 3^2 \cdot 5^2} = 2 \cdot 3 \cdot 5 = 30$$

is an integer. Since 2, 3, and 5 are all primes, we see that $K = 6$ is the smallest possibility that works.

Answer: 6

Problem 10 Solution

Since the checkers are in different rows and columns, the 4 checkers will be placed in the 'top', 'bottom', 'left', and 'right' edges of the chessboard. Further, none of the 4 corners is possible, so we only have to ensure the 'top' and 'bottom' checkers are not in the same column and the 'left' and 'right' checkers are not in the same row. There are 6 columns to choose from for the top checker, which leaves 5 columns for the bottom checker. Hence there are $6 \cdot 5 = 30$ ways to place the top and bottom checkers. Similarly there are 30 ways for the left and right checkers, which leads to $30^2 = 900$ total ways.

Answer: 900

Problem 11 Solution

First note the triangle must have angles $120°, 30°, 30°$. Let x denote the length of one of the remaining sides. Using the Law of Sines we have

$$\frac{x}{\sin(30°)} = \frac{4\sqrt{3}}{\sin(120°)}.$$

Therefore,

$$x = \frac{1}{2} \cdot 4\sqrt{3} \cdot \frac{2}{\sqrt{3}} = 4.$$

The final answer is $2x = 8$.

Answer: 8

Problem 12 Solution

A root must satisfy $7x - 13 = 0$, $x^2 + 7x - 13 = 0$, or $x^2 - 7x + 13 = 0$. The first equation has one real solution, the second equation has two real solutions since the discriminant $7^2 - 4(1)(-13) = 101 > 0$, and the third equation has no real solutions since the discriminant $(-7)^2 - 4(1)(13) = -3 < 0$.

Answer: 3

Problem 13 Solution

Since O and Q are on the x-axis, if $OP = PQ$ then P has x-coordinate

$$\frac{0 + 2\sqrt{2}}{2} = \sqrt{2}.$$

Hence $P = (\sqrt{2}, 2)$ and thus OPQ is a triangle with base $OQ = 2\sqrt{2}$ and height 2. Hence the triangle has area $2\sqrt{2} = \sqrt{8}$ so our answer is 8.

Answer: 8

Problem 14 Solution

If George wears pants, he has

$$4 \cdot 3 \cdot 4 = 48$$

different outfits with a shirt, pants, and shoes. If he wears shorts, there are

$$4 \cdot 5 \cdot 4 \cdot 2 = 160$$

outfits with a shirt, shorts, shoes, and a hat. We then multiply 160 by 2 because he can wear a jacket (or not) with each of the outfits. This gives

$$48 + 2 \cdot 160 = 368$$

outfits in total.

Answer: 368

Problem 15 Solution

Consider three cases: (i) $x < 0$, (ii) $0 \le x < 1$, or (iii) $x \ge 1$. In case (i) we have $-x - 2 = -1 + x$ so $2x = -1$ and $x = -1/2$. In case (ii), $x - 2 = -1 + x$ so $-2 = -1$ which is impossible. Lastly, for (iii), $x - 2 = -x + 1$ so $2x = 3$ and $x = 3/2$. There $m = 3/2$ and $n = -1/2$ so $m - n = 2$.

Answer: 2

Problem 16 Solution

We have $\sin^2(x) + \cos^2(x) = 1$ so $0.36 + \cos^2(x) = 1$ and $\cos^2(x) = 0.64$ so $\cos(x) = \pm 0.8$. However, all angles are $\le 90°$ in a right triangle, so $\cos(x) = 0.8$. Finally, $\cos(180 - x) = -\cos(x) = -0.8$.

Answer: -0.8

Problem 17 Solution

Using laws of logarithms we have

$$\log_2(x) + \log_2(x^2) = \log_2(x) + 2\log_2(x) = 3\log_2(x),$$

so dividing by 3 we need $\log_2(x) = 7$. Hence $x = 2^7 = 128$.

Answer: 128

Problem 18 Solution

Triangle ADE has perimeter 12, so all its sides are length 4. Note that if we add the perimeters of $\triangle ADE$ and trapezoid $DECB$ we would get the perimeter of $\triangle ABC$ except we are counting ED two times instead of none. As ED is length 4, we have the perimeter of ABC is $12 + 16 - 2 \cdot 4 = 20$.

Answer: 20

Problem 19 Solution
Factoring we have

$$\frac{(x+4)(x-2)}{(x+1)(x-2)} = \frac{x+4}{x+1}$$

as long as $x \neq 2$. Hence we can set $x = 2$ to calculate the limit:

$$\frac{2+4}{2+1} = \frac{6}{3} = 2.$$

Answer: 2

Problem 20 Solution
Zeros at the end of a number come from powers of 10. Since $10 = 2 \cdot 5$, we look at pairs the number of 2's and 5's in the prime factorization of 30!. There are a lot more 2's than 5's so we focus on counting how many times 5 is a factor of 30!. Note that the multiples of 5: $5, 10, 15, 20, 25, 30$ each contribute one power of 5 (for 6 powers of 5 so far. However, $25 = 5^2$ so it gives on extra power of 5. Hence there are $6 + 1 = 7$ powers of 5 in the factorization of 30!, so there are 7 zeros.

Answer: 7

2.3 ZIML December 2016 Division H

Below are the solutions from the Division H ZIML Competition held in December 2016.

The problems from the contest are available on p.25.

Problem 1 Solution
The constant term is the product of all the constant terms, so $1 \cdot 2^2 \cdot 2 = 8$.

Answer: 8

Problem 2 Solution
The volume of a cylinder is $\pi r^2 h$ where r is the radius and h is the height. Since 10 inches is $\frac{5}{6}$ of a foot, we have

$$\pi r^2 \cdot \frac{5}{6} = \frac{15\pi}{8},$$

where r is measured in feet. Solving for r we have

$$r^2 = \frac{15}{8} \cdot \frac{6}{5} = \frac{9}{4},$$

so $r = \frac{3}{2}$ feet. Converting to inches we have $r = \frac{3}{2} \cdot 12 = 18$ inches.

Answer: 18

Problem 3 Solution
$75 = 3 \cdot 5^2, 360 = 2^3 \cdot 3^2 \cdot 5$ so $\operatorname{lcm}(75, 360) = 2^3 \cdot 3^2 \cdot 5^2 = 1800$. The smallest multiple of this that is a perfect square is $2^4 \cdot 3^2 \cdot 5^2 = 3600$.

Answer: 3600

Problem 4 Solution

Let $g(x) = -cx^3$ so that $f(x) = g(x) + 12$. We then know $g(2) = f(2) - 12 = 10 - 12 = -2$. Further, $g(x)$ is an odd function, so $g(x) = -g(-x)$ for all x. Therefore $g(-2) = -g(2) = -(-2) = 2$. Thus $f(2) = 2 + 12 = 14$.

Answer: 14

Problem 5 Solution

The smallest possible median is 1, if at least 5 of the elements are 1. If we then make the remaining elements as large as possible (each 10), we get a mean of

$$(1 + 1 + 1 + 1 + 1 + 10 + 10 + 10 + 10) \div 9 = 5.$$

This gives a difference of $5 - 1 = 4$, the largest possible. Note it is also possible to achieve this difference by making the median as large as possible (10) and make the remaining elements 1. In this case the median is 10 and the mean is 6, so the difference is still 4.

Answer: 4

Problem 6 Solution

We have $\sqrt{6b} = 4a = 4(3\sqrt{2}) = 12\sqrt{2}$. Squaring both sides gives $6b = 288$ so $b = 48$.

Answer: 48

Problem 7 Solution

Consider the labeled diagram

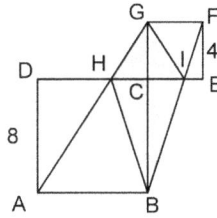

Hence the shaded region is $[ABFG] - [ABH] - [FGI]$. $ABFG$ is a trapezoid, hence has area $(8+4) \cdot 12/2 = 72$. We have $[ABH] = 8 \cdot 8/2 = 32, [FGI] = 4 \cdot 4/2 = 8$ so $[GHBI] = 72 - 32 - 8 = 32$.

Answer: 32

Problem 8 Solution
Completing the square we get $(x+5)^2 + (y-12)^2 = 87 + 25 + 144 = 256$, so the circle has radius $\sqrt{256} = 16$.

Answer: 16

Problem 9 Solution
Isolating the square root we have $\sqrt{2-x} = 4+x$. Squaring both sides gives

$$2 - x = x^2 + 8x + 16 \text{ or } x^2 + 9x + 14 = 0.$$

Factoring we have $(x+2)(x+7) = 0$ so $x = -2$ or $x = -7$. However, plugging in $x = -7$ we get $\sqrt{2-(-7)} - (-7) = \sqrt{9} + 7 = 3 + 7 = 10 \neq 4$, so the only real solution is $x = -2$.

Answer: -2

Problem 10 Solution
There are 6 different rolls possible. Hence there are

$$6 \cdot 6 \cdot 6 \cdot 6 = 6^4$$

total outcomes. If all 4 rolls are different, there are

$$6 \cdot 5 \cdot 4 \cdot 3$$

outcomes (as each roll has one less possibility). Hence the probability is

$$\frac{6 \cdot 5 \cdot 4 \cdot 3}{6^4} = \frac{5}{18},$$

so $N + M = 5 + 18 = 23$.

Answer: 23

Problem 11 Solution

If the discriminant $\Delta > 0$, a quadratic equation has two distinct real roots. As

$$\Delta = 8^2 - 4 \cdot 4m = 64 - 16m > 0$$

we must have $m < 4$. Hence $N = 4$.

Answer: 4

Problem 12 Solution

It is possible for $30°$ to be either the angles repeated twice or not, leading to two different answers. If there are two $30°$ angles, the remaining angle is $180 - 30 - 30 = 120°$. Otherwise two equal angles must add up to $180 - 30 = 150°$, hence are both $150° \div 2 = 75°$. Thus the largest possible is $120°$.

Answer: 120

Problem 13 Solution

Since the lines are parallel, they have the same slope. Hence the slope of both lines is

$$\frac{1 - 0}{0 - (-4)} = \frac{1}{4}.$$

Therefore using slope intercept form, line m has equation

$$y = \frac{1}{4} \cdot x - 3.$$

Since $(t, 0)$ is on this line, we must have

$$0 = \frac{1}{4} \cdot t - 3$$

so we can solve for $t = 12$.

Answer: 12

Problem 14 Solution

Let $u = \sin(x)$, so our equation becomes

$$4u^2 + u - 3 = 0.$$

Factoring we have

$$(4u - 3)(u + 1) = 0$$

so $u = \frac{3}{4} = 0.75$ or $u = -1$. If $\sin(x) = -1$, then x is not between $0°$ and $90°$, hence $\sin(x) = u = 0.75$.

Answer: 0.75

Problem 15 Solution

Note that the palindrome is determined if we know the first 4 letters (as they must be the same as the first 3 letters in reverse). There are no restrictions on these 4 letters, so we have 4 choices for each. Hence there are $4 \cdot 4 \cdot 4 \cdot 4 = 256$ palindromic oligomers.

Answer: 256

Problem 16 Solution

We know $\overline{AB} \parallel \overline{CD}$ because $ABCD$ is a trapezoid. Therefore $\triangle ABE$ is similar to $\triangle CDE$ as they share the same three angles.

Let $DE = x$, so

$$\frac{DE}{AE} = \frac{CD}{AB} \Rightarrow \frac{x}{x+3} = \frac{4}{8}.$$

Solving for x we have $2x = x+3$ so $x = 3$. An identical calculation shows $CE = 2$. Therefore the perimeter of $\triangle CDE$ is $2+3+4 = 9$.

Answer: 9

Problem 17 Solution

Suppose that x is not an integer. Then $\lfloor x \rfloor = N$ for some integer N and $\lceil x \rceil = N+1$ (for that same integer). Thus

$$f(x) = 5\lfloor x \rfloor - 3\lceil x \rceil = 5(N) - 3(N+1) = 2N - 3,$$

so $A + B = 2 - 3 = -1$.

Answer: -1

Problem 18 Solution

The number of friends must be a factor of $99 - 8 = 7 \cdot 13$. Since there are more than 7 friends, the answer must be 13 or 91. However, if there are 91 friends, not every friend gets 2 jellybeans. Hence there must be 13 friends.

Answer: 13

Problem 19 Solution

Note that 12 has 6 factors, which can be paired so that

$$(1 \cdot 12) \cdot (2 \cdot 6) \cdot (3 \cdot 4) = 12^3.$$

With this reasoning, we see that any number with exactly 6 factors will have the property we want. Then note that 18 is the next number with exactly 6 factors $(1,2,3,6,9,18)$, so the answer is 18.

Answer: 18

Problem 20 Solution

We have

$$\frac{2+\cos(x)}{1+2\sec(x)} \cdot \frac{\cos(x)}{\cos(x)} = \frac{(2+\cos(x))\cdot\cos(x)}{\cos(x)+2} = \cos(x).$$

Hence $A = 1, B = 0$ so $A + B = 1$.

Answer: 1

2.4 ZIML January 2017 Division H

Below are the solutions from the Division H ZIML Competition held in January 2017.

The problems from the contest are available on p.31.

Problem 1 Solution

The time the maximum occurs at the vertex of the quadratic function, which is

$$t = -\frac{b}{2a} = -\frac{64}{2(-16)} = 2.$$

Hence the maximum height is

$$h = 3 + 64(2) - 16(2^2) = 3 + 128 - 64 = 67.$$

Answer: 67

Problem 2 Solution

Let x be the number of tacos sold so there were $155 - x$ burritos sold. We have the equation

$$3x + 8.50(155 - x) = 729$$

so solving for x we see three were 107 tacos sold.

Answer: 107

Problem 3 Solution

Using the distance formula, we calculate the radius of the circle is

$$\sqrt{(-3-2)^2 + (4-2)^2} = \sqrt{25+4} = \sqrt{29}.$$

Thus the equation of the circle is

$$(x-2)^2 + (y-2)^2 = 29.$$

Expanding and rearranging we get

$$x^2 + y^2 = 4x + 4y + 21,$$

so $C = 21$.

Answer: 21

Problem 4 Solution

Let h be the height of the trapezoid, so

$$\frac{5+21}{2} \cdot h = 78$$

and we see that $h = 6$. Dividing the trapezoid into a rectangle and two right triangles gives the following diagram:

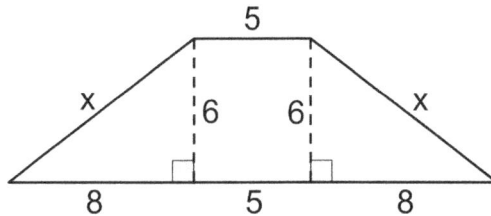

Using the Pythagorean theorem we can solve to get $x = 10$, so the perimeter is

$$21 + 10 + 5 + 10 = 46.$$

Answer: 46

Problem 5 Solution

The number of dollars is a multiple of $\text{lcm}(24, 40) = 120$, so there are $120k$ dollars, where k is a positive integer. Based on the description, there must be $120k \div 24 = 5k$ nieces and $120k \div 40 = 3k$ nephews. It follows that each niece or nephew gets $120k \div (5k + 3k) = 15$ dollars.

Answer: 15

Problem 6 Solution

The first ticket can be given to 1 of 10 people, so there are 10 options for the first ticket. After the first ticket has been distributed, there are 9 ways that the second ticket can be given. Repeating this process, we observe that there are

$$\frac{10!}{6!} = 10 \times 9 \times 8 \times 7 = 5040$$

ways to distribute the 4 tickets among 10 friends.

Answer: 5040

Problem 7 Solution

Since $\angle ABC = 90°$, we have $\overset{\frown}{CA} = 180°$. Hence, $\overset{\frown}{BC} : \overset{\frown}{AB} = 4 : 5$. Hence, $\overset{\frown}{BC} = 80°$ and $\angle BAC = 80/2 = 40°$.

Answer: 40

Problem 8 Solution

Note $2592 = 2^5 \cdot 3^4$. To form a factor that is a perfect square, we use even powers of $2, 3 : 2^0, 2^2, 2^4, 3^0, 3^2, 3^4$. There are $3 \cdot 3 = 9$ combinations of these which give the perfect squares as factors.

Answer: 9

Problem 9 Solution

Rearranging the inequality, it is equivalent to

$$-2 \geq x.$$

Hence there are 4 integers $(-2, -3, -4, -5)$ in the desired range that satisfy the inequality.

Answer: 4

Problem 10 Solution

Note $\overline{EF}, \overline{GH}$ divide $ABCD$ into four parallelograms, whose areas are proportional. That is $[AEPG]/[GPFD] = [EBHP]/[PHCF]$ so $[AEPG] = 10 \cdot 16/8 = 20$. Hence the total area of $ABCD$ is 54.

Answer: 54

Problem 11 Solution

Using the rules of exponents we know

$$x^{a^2-b^2} = x^{28} \text{ and } x^{b-a} = x^7,$$

so $a^2 - b^2 = 28$ and $b - a = 7$. Factoring we have

$$28 = a^2 - b^2 = (a-b)(a+b) = -7(a+b)$$

so $a + b$ must be -4.

Answer: -4

Problem 12 Solution

The internal angles of a regular pentagon are

$$\frac{180(5-2)}{5} = 108°$$

so the external angles are $180° - 108° = 72°$. Hence the difference is

$$108° - 72° = 36°.$$

Answer: 36

Problem 13 Solution

We have

$$\log_{\sqrt{3}}(x) = \log_3(x)/\log_3(\sqrt{3}) = 2\log_3(x) = \log_3(x^2)$$

and similar methods give

$$2\log_3(x) = \log_3(x^2) \text{ and } -\log_9(x) = \log_3(x^{-1/2}).$$

Combining gives $\log_3(x^{7/2})$ so $p+q=9$.

Answer: 9

Problem 14 Solution

Since the size of the regions are proportional let x be the probability the spinner lands in the '1' region, so the other probabilities are $2x$, $3x$, $4x$, and $5x$ respectively. Probabilities add up to 1, so

$$x + 2x + 3x + 4x + 5x = 15x = 1 \Rightarrow x = \frac{1}{15}.$$

Hence the probability it lands in the '4' region is

$$\frac{4}{15}$$

so $M - N = 15 - 4 = 11$.

Answer: 11

Problem 15 Solution

Using the Law of Cosines we calculate

$$AC^2 = 13^2 + 15^2 - 2 \cdot 13 \cdot 15 \cdot \frac{33}{65} = 196,$$

so $AC = 14$. Using the Law of Sines we have $\dfrac{\sin C}{15} = \dfrac{\sin B}{14}$ so $\sin C = \dfrac{15\sin B}{14}$. Lastly,

$$\sin^2 B = 1 - \frac{33^2}{65^2} = \frac{65^2 - 33^2}{65^2} = \frac{(65-33)(65+33)}{65^2} = \frac{56^2}{65^2},$$

so $\sin C = \dfrac{15}{14} \cdot \dfrac{56}{65} = \dfrac{12}{13}$ and $P + Q = 12 + 13 = 25$.

Answer: 25

Problem 16 Solution

Using symmetry we look at spheres with centers across the diagonal:

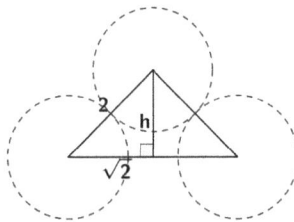

Let $h + 1$ denote the height of the center of the fifth sphere off the ground. Setting up right triangles gives us $2^2 = 2 + h^2$, so solving for h gives $\sqrt{2}$ so $L = 2$.

Answer: 2

Problem 17 Solution

$i^2 = -1$, so $i^3 = -i$, $i^4 = 1$, and $i^5 = i$. We have

$$f(i) = 6i^5 + 5i^4 + 4i^3 + 3i^2 + 2i + 1 = 6i + 5 - 4i - 3 + 2i + 1 = 3 + 4i$$

so $A^2 + B^2 = 3^2 + 4^2 = 25$.

Answer: 25

Problem 18 Solution

Note that counting means is the same as counting sums. The smallest sum is 5 and the largest sum is 50. Every sum in between is possible, so there are $50 - 5 + 1$ different sums and so $50 - 5 + 1$

different means.

Answer: 46

Problem 19 Solution

Recall every number has a prime factorization. Since the small-est primes larger than 23 are $29, 31, 33, 37$. Hence the small-est possible composite numbers not deleted are $29^2, 29 \cdot 31, 29 \cdot 33, 31 \cdot 31, 31 \cdot 31$, etc. The only ones smaller than 1000 are $29^2 = 841, 29 \cdot 31 = 899, 31 \cdot 31 = 961$ with sum $841 + 899 + 961 = 2701$.

Answer: 2701

Problem 20 Solution

Note $\log_4(x) = \log_2(x)/2$ so the equation is

$$\log_2^2(x) + 2\log_2(x) - 3 = 0.$$

Let $z = \log_2(x)$ so we have

$$z^2 + 2z - 3 = (z+3)(z-1) = 0$$

or $z = -3, z = 1$. If $z = 1$ we have $x = 2$ and if $z = -3$ we have $x = 1/8$. Both are solutions. Hence the product is 0.25.

Answer: 0.25

2.5 ZIML February 2017 Division H

Below are the solutions from the Division H ZIML Competition held in February 2017.

The problems from the contest are available on p.37.

Problem 1 Solution

Alice got the correct line of $y = 3x - 2$ (in slope-intercept form). Bob mixed up so he got $y = -2x + 3$. Setting these the y values equal gives $3x - 2 = -2x + 3$ so $x = 1$ after combining like terms. Plugging back in gives $y = -2(1) + 3 = 1$, so they intersect when $y = 1$.

Answer: 1

Problem 2 Solution

Note completing the square we have that

$$f(x) = x^2 - 4x + 4 - 3 = (x-2)^2 - 3.$$

Hence $f(x)$ is a parabola opening upwards with vertex $(2, -3)$. The maximum value for $|x| \le 5$ therefore occurs when $x = -5$. Plugging this in we have a maximum value of

$$f(-5) = (-5 - 2)^2 - 3 = 49 - 3 = 46.$$

Answer: 46

Problem 3 Solution

Let the height be x. The area A minus the area B is equal to the semicircle minus the triangle

$$\frac{10^2 \pi}{2} - \frac{20x}{2} = 50\pi - 100.$$

Solving for x we get $x = 10$.

Answer: 10

Problem 4 Solution

Given that there are 50 students that scored above 600 on either the critical reading or math section of the SAT, the overlap of students that score above 600 on either sections represent the number of students that score above 600 on both sections.

Therefore, there are

$$(35 + 25) - 50 = 10$$

students that score above 600 on both critical reading and math sections of the SAT.

The maximum total score that these 10 students can score is

$$800 + 800 = 1600,$$

and the maximum total score that the remaining 40 students can score is

$$800 + 600 = 1400.$$

Therefore, the sum of the 50 SAT total scores is

$$40 \times 1400 + 10 \times 1600 = 72000$$

and the average of the 50 SAT total scores is

$$72000 \div 50 = 1440.$$

Answer: 1440

Problem 5 Solution

Suppose x is an integer. Therefore, we can get rid of the floor functions and solve the following equation: $x^2 + x + 1 = 5x - 2$. Equivalently, $x^2 - 4x + 3 = 0$, so $(x-1)(x-3) = 0$. The solutions are $x = 1$ and $x = 3$.

Now suppose that x is not an integer. Therefore, let $x = \lfloor x \rfloor + \{x\}$. Plugging this in for x yields,

$$\lfloor x \rfloor^2 + \lfloor x \rfloor + 1 = 5(\lfloor x \rfloor + \{x\}) - 2$$

or equivalently,

$$\lfloor x \rfloor^2 - 4\lfloor x \rfloor + 3 = 5\{x\}.$$

This implies that $5\{x\}$ must be an integer, so $\{x\} = \frac{k}{5}$ for $k = 1, 2, 3, 4$. Therefore, since

$$\lfloor x \rfloor^2 - 4\lfloor x \rfloor + 3 = k$$

and $\lfloor x \rfloor$ is an integer, $k = 3$. Therefore, $\lfloor x \rfloor = 0$ or $\lfloor x \rfloor = 4$ and $x = \frac{3}{5}$ and $x = \frac{23}{5}$.

Therefore, the answer is $\frac{3}{5} + \frac{23}{5} + 1 + 3 = 9.2$.

Answer: 9.2

Problem 6 Solution

Note we have $\angle A = 180° - 45° - 75° = 60°$. We have $\sin 60° = \sqrt{3}/2, \sin 45° = \sqrt{2}/2$ so

$$\frac{\sin(A)}{a} = \frac{\sin(B)}{b} \Rightarrow \frac{\sqrt{3}/2}{a} = \frac{\sqrt{2}/2}{4}.$$

Cross multiplying and simplifying we have

$$a = \frac{4\sqrt{3}}{\sqrt{2}} = 2\sqrt{6}$$

so $R + S = 8$.

Answer: 8

Problem 7 Solution
Recall the shortest distance is perpendicular to the line. Hence we find where $y = 2x + 5$ intersects the line with slope $-1/2$ containing $(0,0)$ which is

$$y = -\frac{x}{2}.$$

Calculating where these two lines intersect gives the point $(-2, 1)$. Hence the distance is

$$\sqrt{(-2-0)^2 + (1-0)^2} = \sqrt{5}$$

so $D = 5$.

Answer: 5

Problem 8 Solution
The new box (with volume 24 cubic feet) can have dimensions

$$4 \times 1 \times 6, \quad 2 \times 2 \times 6, \quad \text{or} \quad 2 \times 1 \times 12$$

These new boxes have surface area

$$2(4 \cdot 1 + 4 \cdot 6 + 1 \cdot 6) = 68,$$

$$2(2 \cdot 2 + 2 \cdot 6 + 2 \cdot 6) = 56,$$

and

$$2(2 \cdot 1 + 2 \cdot 12 + 1 \cdot 12) = 76.$$

The sum of these is $68 + 56 + 76 = 200$.

Answer: 200

Problem 9 Solution

$\sqrt{x+3} = \sqrt{3x+2} - 1$, squaring, $x+3 = 3x - 2 - 2\sqrt{3x-2} + 1$, thus $x - 2 = \sqrt{3x-2}$, square again, $x^2 - 7x + 6 = 0$, and get $x = 1$ or 6. Double checking we see that $x = 1$ is extraneous, but $x = 6$ works. Hence the sum of all real solutions is just 6.

Answer: 6

Problem 10 Solution

Given the conditions of the problem, the 5 people must sit in seats $1, 3, 5, 7, 9$.

Since these seats are required to be filled, we determine the number of ways to permute 5 people. This is done in

$$5 \times 4 \times 3 \times 2 \times 1 = 120$$

ways.

Answer: 120

Problem 11 Solution

Recall $(5, 12, 13)$ is a Pythagorean triple. As $\frac{\pi}{2} < \theta < \pi$, we have $\cos(\theta) < 0$ and hence

$$\tan(\theta) = -\frac{5}{12}.$$

This gives

$$\tan(2x) = \frac{2\tan(x)}{1 - \tan^2(x)} = \frac{-120}{119}.$$

Calculating $M - N = 119 - (-120) = 239$.

Answer: 239

Problem 12 Solution

Note that cell n will be left open if n has an odd number of factors. Hence, only cells whose number is a square will be unlocked.

$31^2 = 961$ but $32^2 = 1024$, so there are 31 perfect squares less than 1000.

Answer: 31

Problem 13 Solution

Using the change of base formula we can rewrite the equation as

$$2\log_3 x + \log_3 x + \frac{1}{2}\log_3 x = -\frac{21}{2} \text{ so } \log_3(x) = -3.$$

Hence $x = 3^{-3} = 1/27$. This is the only, hence smallest, positive solution, so $P + Q = 1 + 27 = 28$.

Answer: 28

Problem 14 Solution

The two chords divide each other in segments of 4 and 8 inches. The perpendicular bisector of each chord goes through the center of the circle, so we can form a right triangle with sides 2 and 6 with hypotenuse r, the radius of the circle. Hence, $r = \sqrt{2^2 + 6^2} = \sqrt{40}$. The circle thus has area 40π and $L = 40$.

Answer: 40

Problem 15 Solution

We have $f(7) = 3 \cdot 7 + 2 = 23$. If $f^{-1}(x) = y$, we have

$$3y + 2 = -7 \Rightarrow y = -3.$$

Hence $f(7)f^{-1}(-7) = 23 \cdot -3 = -69$.

Answer: -69

Problem 16 Solution

Suppose the sides have length $3k, 4k, 5k$ (so the hypotenuse has length $5k$). Hence the triangle has area $\frac{1}{2} \cdot 12k^2$. If we let h denote the length of the altitude to the hypotenuse we also have the area is $\frac{1}{2} \cdot 5k \cdot h$. Hence we must have

$$\frac{1}{2} \cdot 12k^2 = \frac{1}{2} \cdot 5k \cdot h \Rightarrow h = \frac{12}{5} \cdot k.$$

For the smallest k for which h is an integer is thus $k = 5$. Hence the triangle has side lengths $15, 20, 25$ and thus area

$$\frac{1}{2} \cdot 15 \cdot 20 = 150.$$

Answer: 150

Problem 17 Solution

Recall $(a+b)^4 = a^4 + 4a^3 b + 6a^2 b^2 + 4ab^3 + b^4$ so

$$\begin{aligned}(2+i)^4 &= 2^4 + 4 \cdot 2^3 \cdot i + 6 \cdot 2^2 \cdot i^2 + 4 \cdot 2 \cdot i^3 + i^4 \\ &= 16 + 32i - 24 - 8i + 1 \\ &= -7 + 24i\end{aligned}$$

Thus,
$$S^2 + T^2 = 7^2 + 24^2 = 25^2 = 625.$$

Alternatively, note $|2+i| = \sqrt{2^2 + 1^2} = 5$. Therefore

$$|(2+i)^4| = |(2+i)|^4 = \sqrt{5}^4 = 25.$$

As $S^2 + T^2$ is $|(2+i)^4|^2$, we have $S^2 + T^2 = 25^2 = 625$.

Answer: 625

Problem 18 Solution

Note the only possibilities are you rolled a 5 and got all 5 heads, or rolled a 6 and got 5 heads and 1 tail. The first case has probability

$$\frac{1}{6} \cdot \left(\frac{1}{2}\right)^5 = \frac{1}{192}.$$

In the second case, note there are 6 possibilities for when in the 6 flips you get a tails, so the second case has probability

$$\frac{1}{6} \cdot 6 \cdot \left(\frac{1}{2}\right)^6 = \frac{1}{64}.$$

Adding we get a final probability of

$$\frac{1}{192} + \frac{1}{64} = \frac{8}{384} = \frac{1}{48},$$

so our answer is $M = 48$.

Answer: 48

Problem 19 Solution

Factoring the given equation we have

$$(a+b)^2 - c^2 = 3ab \text{ so } c^2 = a^2 + b^2 - ab.$$

Therefore, according to the Law of Cosines, $\cos \angle C = \frac{1}{2}$ and hence $\angle C = 60°$.

Answer: 60

Problem 20 Solution

When we square a number, the last digit depends only on the last digit of the number we are squaring. For example the lat digits of 1^2, 11^2, 21^2, etc. are all 1. Thus, the last digit of the sum $1^2 + 2^2 + 3^2 + \cdots + 99^2$ is the same as the last digit of the sum

$$\underbrace{1^2 + \cdots + 1^2}_{10 \text{ times}} + \underbrace{2^2 + \cdots + 2^2}_{10 \text{ times}} + \cdots + \underbrace{9^2 + \cdots + 9^2}_{10 \text{ times}}$$

that is, $10 \times (1^2 \cdots + 9^2)$. Therefore, the last digit of the sum is 0.

Answer: 0

2.6 ZIML March 2017 Division H

Below are the solutions from the Division H ZIML Competition held in March 2017.
The problems from the contest are available on p.43.

Problem 1 Solution
Note that $s + 6t = (3s + 5t) - (2s - t) = 10 - 7 = 3$. Therefore, $\frac{1}{2}s + 3t = \frac{3}{2}$

Answer: 1.5

Problem 2 Solution
If the base length is 6 and the area of $ABCD$ is 24, then the height is $24 \div 6 = 4$. Since the height of the parallelogram divides the base into two equal parts, this forms a right triangle with leg lengths 3 and 4 and therefore the hypotenuse has length 5.

The perimeter of the parallelogram is thus $6 + 5 + 6 + 5 = 22$.

Answer: 22

Problem 3 Solution
Note that $f(x) = \frac{1}{3}x - 3$ implies that $f^{-1}(x) = 3(x + 3)$. If $f(x)f^{-1}(x) = 0$, then $f(x) = 0$ or $f^{-1}(x) = 0$. Therefore, $x = 9$ and $x = -3$ respectively. Therefore, $9 + (-3) = 6$.

Answer: 6

Problem 4 Solution
Note that the only ways to obtain a product greater than or equal to 24 is $(6,4), (4,6), (5,5), (6,5)(5,6)$, and $(6,6)$. As there are $6^2 = 36$ total outcomes, the probability is

$$\frac{6}{36} = \frac{1}{6}.$$

hence $p + q = 1 + 6 = 7$.

Answer: 7

Problem 5 Solution

Note that the distance from the coordinate to the origin is $\sqrt{3^2 + 4^2} = 5$. The circle $x^2 + y^2 = 9$ has radius 3. Therefore, the shortest length from the coordinate $(3,4)$ to the circle defined as $x^2 + y^2 = 9$ is $5 - 3 = 2$.

Answer: 2

Problem 6 Solution

Factoring both sides of the equation we get

$$2^x(1 + 2^x + 2^{2x}) = 2^3 \cdot 73.$$

Since 2 is not a factor of 73 and 2 is not a factor of $1 + 2^x + 2^{2x}$, we must have $2^x = 2^3$ or $x = 3$. To check, note that $1 + 2^3 + 2^6 = 73$.

Answer: 3

Problem 7 Solution

First note that the ones digit has to be a 5 or 6. If the number is $10x + 5$, then

$$(10x + 5)^2 = 100x^2 + 100x + 25,$$

so the number must be 25.

If the number is $10x + 6$, then

$$(10x + 6)^2 = 100x^2 + 120x + 36 = 100(x^2 + x) + 20x + 36.$$

Therefore the new number's last two digits are determined by $20x + 36$. Hence we must have

$$20x + 36 = 10x + 6 \text{ or } 10x + 106 \text{ or } 10x + 206 \text{ or } \ldots.$$

so
$$x = -3 \text{ or } 7 \text{ or } 17 \text{ or } \dots$$

Here the only answer that makes sense is $x = 7$ and thus the only number in this case is 76.

Therefore the sum of the possibilities is $25 + 76 = 101$.

Answer: 101

Problem 8 Solution

The triangle is a right isosceles triangle, so we have a semicircle of radius 5 plus $45/360 = 1/8$th of a circle of radius 10 minus an isosceles right triangle with legs 10:

$$\frac{1}{2}\pi \times 5^2 + \frac{1}{8}\pi \times 10^2 - \frac{1}{2} \times 10^2 = 25\pi - 50.$$

Hence $R + S = 25 + (-50) = -25$.

Answer: -25

Problem 9 Solution

Note that counting means is the same as counting sums. The smallest sum is 5 and the largest sum is 50. Every sum in between is possible, so there are $50 - 5 + 1$ different sums and hence $50 - 5 + 1$ different means.

Answer: 46

Problem 10 Solution

Since Megan runs twice as fast as Sally, the distance she runs will be twice as long as Sally's. Let's denote the distance Sally ran by x.

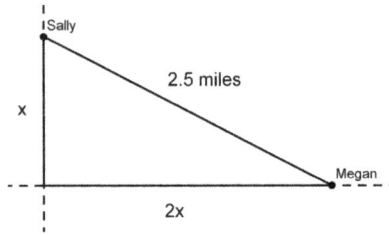

We have then a right triangle where the ratios of the sides are $1 : 2 : \sqrt{5}$. Thus, Sally ran

$$x = \frac{2.5}{\sqrt{5}} = \frac{\sqrt{5}}{2}$$

and Megan ran

$$2x = \sqrt{5}$$

miles in 10 minutes. Hence Megan's speed is $6\sqrt{5}$ miles per hour.

Answer: 11

Problem 11 Solution

Break into cases based on whether $x \geq -\frac{1}{2}$ or $x < -\frac{1}{2}$. If $x \geq -\frac{1}{2}$,

$$|x - |2x + 1|| = |x + 1| = 3,$$

so $x = 2$ or $x = -4$. As $-4 < -\frac{1}{2}$, $x = 2$ is the only solution for this case.

If $x < -\frac{1}{2}$,

$$|x - |2x + 1|| = |3x + 1| = 3,$$

then $x = 2/3$ or $x = -4/3$. Here only $-4/3$ works.

This gives a final solution of $-4/3 + 2 = 2/3 \approx 0.67$.

Answer: 0.67

Problem 12 Solution

We can use the Law of Cosines to find out the angle. We have that

$$\left(\sqrt{90}\right)^2 = \left(\sqrt{18}\right)^2 + 6^2 - 2 \cdot \sqrt{18} \cdot 6 \cdot \cos\theta.$$

Simplifying

$$90 = 18 + 36 - 36\sqrt{2}\cos\theta$$

hence

$$\cos\theta = -\frac{\sqrt{2}}{2},$$

which means $\theta = 135°$.

Answer: 135

Problem 13 Solution

The quadratic equation has two complex roots, so they must be conjugates of each other. Hence we get that

$$2 = b \text{ and } a = -1.$$

Therefore

$$x^2 + px + q = (x - (2 - i))(x - (2 + i)) = x^2 - 4x + 5,$$

so $q = 5$.

Answer: 5

Problem 14 Solution

Note that $396 = 2^2 \cdot 3^2 \cdot 11$, so exactly one of A or B is divisible by 11. Since $\gcd(B,C) = 33 = 3 \cdots 11$, it must be the case that B is divisible by 11. Therefore, $\gcd(11A, B) = 11 \times \gcd(A, B) = 132$.

Answer: 132

Problem 15 Solution

First note the triangle must have angles $120, 30, 30$. Hence the other two sides have the same length. Let x denote the length of one of the remaining sides. Using the Law of Sines we have

$$\frac{x}{\sin(30°)} = \frac{4\sqrt{3}}{\sin(120°)}. \text{ Therefore, } x = \frac{1}{2} \cdot 4\sqrt{3} \cdot \frac{2}{\sqrt{3}} = 4. \text{ Thus}$$

our final answer is $4 + 4 = 8$.

Answer: 8

Problem 16 Solution

Note that $\log_x(4x - 3) = 2$ implies that $4x - 3 = x^2$ or $x^2 - 4x + 3 = (x - 3)(x - 1) = 0$. This implies that $x = 3$ or $x = 1$. However, $x \neq 1$ since the base of a logarithm cannot be 1. Therefore, $x = 3$.

Answer: 3

Problem 17 Solution

Let $x = 2017$, then the expression is

$$\frac{(x^2 - x - 6)(x^2 + 2x - 3)(x + 1)}{(x - 3)(x - 1)(x + 1)(x + 3)} =$$
$$= \frac{(x + 2)(x - 3)(x + 3)(x - 1)(x + 1)}{(x - 3)(x - 1)(x + 1)(x + 3)}$$
$$= x + 2 = 2019.$$

Answer: 2019

Problem 18 Solution

Given that 7 of the 12 members are seniors and that we require 5 to be in the committee, there are $\binom{7}{5} = 21$ ways to choose the 5 seniors to be members of the committee. The remaining 5 members are not seniors and we require 3 to be in the committee. Therefore, there are $\binom{5}{3} = 10$.

Therefore, there are $21 \times 10 = 210$ ways that this can be done.

Answer: 210

Problem 19 Solution

First note that using the volume formula for the cone, there is

$$\frac{\pi}{3} \times 2^2 \times 4 = \frac{16\pi}{3}$$

cubic inches of ice cream before any is eaten. After you've eaten some ice cream, consider the following side view (where ABC is the original cone, and ADE is the 'half cone').

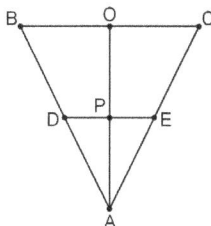

Note that $\triangle AOC \sim \triangle APE$ as they are both right triangles that share angle $\angle OAC$. Since $OA = 4, PA = 2$, we have that the ratio of sides is 2, so

$$PE = \frac{1}{2}OC = 1.$$

Hence the new cone has radius 1, and thus volume

$$\frac{\pi}{3} \times 1^2 \times 2 = \frac{2\pi}{3}$$

cubic inches of ice cream. Hence you are left with 1/8th of the ice cream you started with, so you have eaten 7/8th of the ice cream. As a percentage this is 87.5%.

Answer: 87.5

Problem 20 Solution

Using the double angle formula we can rewrite everything in terms of $\sin(x), \cos(x)$:

$$2\sin(x)\cos(x) = \frac{\sin(x)}{\cos(x)}.$$

Hence either $\sin(x) = 0$, so $x = 0°, 180°, \ldots$, or $\cos^2(x) = \frac{1}{2}$ so $\cos(x) = \pm\frac{\sqrt{2}}{2}$. This gives $x = 45°, 135°, 225°, \ldots$. Therefore $225°$ is the smallest solution with $x > 180°$.

Answer: 225

2.7 ZIML April 2017 Division H

Below are the solutions from the Division H ZIML Competition held in April 2017.

The problems from the contest are available on p.49.

Problem 1 Solution

We need to solve $2 + 3x = x^2 + x$ or $x^2 - 2x - 2 = 0$. Using the quadratic formula we see that $x = 1 \pm \sqrt{3}$. $1 - \sqrt{3} < 0$, so does not work. Hence

$$x = 1 + \sqrt{3} \approx 1 + 1.73 \approx 2.7$$

weeks is the answer.

Answer: 2.7

Problem 2 Solution

We have $\angle APD = 90° - 60° = 30°$. As $\triangle APD$ is isosceles we therefore have $\angle PAD = (180° - 30°)/2 = 75°$.

Answer: 75

Problem 3 Solution

Let x be the number of chickens and y the number of rabbits. Hence counting legs on Sunday we have $2x + 4y = 500$. On Monday he ends up with x old chickens and y new chickens. On Tuesday he ends up with y chickens and $1.5x$ rabbits, so counting legs again on Wednesday we have $2y + 4(1.5x) = 500$ or equivalently $6x + 2y = 500$. Doubling this equation and subtracting it from the first we have $-10x = -500$ so $x = 50$. We can then solve for $y = 100$.

Answer: 50

Problem 4 Solution

Using the Angle Bisector theorem, we have

$$\frac{AB}{AC} = \frac{BE}{EC} = \frac{4}{7}, \quad \frac{AD}{AC} = \frac{DF}{FC} = \frac{9}{7},$$

thus $AB = 8$ and $AD = 16$. Therefore the perimeter of $ABCD$ is

$$AB + BC + CD + DA = 8 + 4 + 7 + 7 + 9 + 16 = 51.$$

Answer: 51

Problem 5 Solution

The sum of Jamie's scores is:

$$70 + 75 + 75 + 85 + 85 + 90 = 480.$$

Let n be the number of quizzes Jamie takes. Assuming that she scores 100 on each quiz she takes, the sum of all Jamie's scores will be $480 + 100n$. Jamie has taken $n + 6$ quizzes. Therefore, Jamie's new quiz average must satisfy is

$$\frac{480 + 100n}{n + 6} \geq 90$$

Solving for n, we will see that

$$480 + 100n \geq 90(n + 6) = 90n + 540$$

or,
$$10n \geq 60.$$

Therefore the least amount of quizzes that Jamie must take is 6.

Answer: 6

Problem 6 Solution

Note when multiplying two numbers, the units digit of the result only depends on the units digit of the numbers being multiplied.

The pattern of the units digit is $4,6,4,6,\ldots$ for powers of 14 and $6,6,6,6,\ldots$ for 16. The exponents are both even, thus the units digits are both 6, and the units digit of the sum $6+6$ is 2.

Answer: 2

Problem 7 Solution

Any quadratic has a line of symmetry through its vertex, so since the quadratic contains $(0,8)$ and $(-6,8)$, the vertex must be $(-3,5)$ (the minimum value always occurs at the vertex). Thus we know the quadratic has formula

$$y = D(x+3)^2 + 5$$

for some number D. Plugging in $x = 0$, $y = 8$ we have

$$8 = D(0+3)^2 + 5 \Rightarrow D = \frac{1}{3}.$$

Hence

$$y = \frac{1}{3}(x+3)^2 + 5 = \frac{1}{3}x^2 + 2x + 8.$$

Thus $A + B + C \approx 0.33 + 2 + 8 = 10.33$.

Answer: 10.33

Problem 8 Solution

Label the lines: (i) $y = x$, (ii) $y = 6 - x$, and (iii) $y = \frac{1}{2}x$. Note (i) and (ii) intersect at $(3,3)$, (i) and (iii) at $(0,0)$, and (ii) and (iii) at $(4,2)$. Further, as the slopes of (i) and (ii) are opposite reciprocals, this triangle is a right triangle, as shown below.

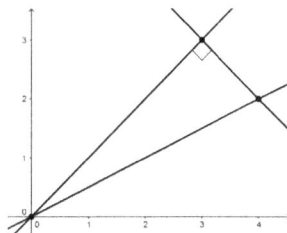

The distance from $(0,0)$ to $(3,3)$ is $\sqrt{3^2 + 3^2} = 3\sqrt{2}$ and the distance from $(3,3)$ to $(4,2)$ is $\sqrt{1^2 + 1^2} = \sqrt{2}$. Hence the triangle has area $\frac{1}{2} \cdot 3\sqrt{2} \cdot \sqrt{2} = 3$.

Answer: 3

Problem 9 Solution

The function $f(x)$ can be written as $f(x) = x^2 - 4$ when $|x| \geq 2$ and $f(x) = 4 - x^2$ for $|x| < 2$, giving the graph shown below:

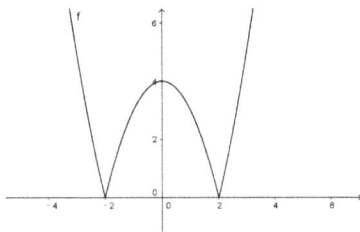

Recall a function must pass the horizontal line test to have an inverse, so if we break the function into the intervals: (i) $x < -2$, (ii) $-2 \leq x < 0$, (iii) $0 \leq x < 2$, and (iv) $x \geq 2$, each interval has an inverse. Further, it is not possible to divide the function into 3 pieces, so our final answer is 4.

Answer: 4

Problem 10 Solution

We have
$$\frac{\sin(30°)}{5} = \frac{\sin(45°)}{b}$$
or, after rearranging,

$$b = 5 \cdot \frac{\sin(45°)}{\sin(30°)} = 5 \cdot \frac{\sqrt{2}/2}{1/2} = 5\sqrt{2} \approx 7.07 \approx 7.1.$$

Answer: 7.1

Problem 11 Solution

Only numbers that are perfect squares have an odd number of factors. (Look at pairing factors, if n is a perfect square, all the other factors can be paired up leaving $\sqrt{n}$ by itself.) Therefore the numbers less than 200 (since $14^2 = 196$) with an odd number of factors will be $1^2, 2^2, 3^2, \ldots, 14^2$, a total of 14 numbers.

Answer: 14

Problem 12 Solution

The volume of a square pyramid can be determined by the following calculation:

$$V = \frac{1}{3} \times 6 \times 6 \times 4 = 48.$$

For the surface area, note that the base is a square and the 4 sides are congruent triangles. These triangles each have a base of length 6. Since the center of the base is 3 away from the sides and the height of the pyramid is 4, we can use the Pythagorean theorem to calculate the height of the triangular sides as $\sqrt{3^2 + 4^2} = \sqrt{25} = 5$. Hence the surface area is

$$S = 4 \cdot \left(\frac{1}{2} \cdot 6 \cdot 5 \right) + 6^2 = 60 + 36 = 96.$$

Therefore $V - S = 48 - 96 = -48$.

Answer: -48

Problem 13 Solution

Pretend that all of the letters are distinct. There are

$$7! = 7 \times 6 \times 5 \times 4 \times 3 \times 2 \times 1 = 840$$

ways to rearrange 7 letters to form distinct words.

Since there are 3 A's and 2 N's in *BANANAS*, we need to rid the duplicates by dividing the total number of ways to rearrange 7 different letters by $3! \times 2! = 6 \times 2 = 12$.

Therefore, the answer is

$$\frac{7!}{3! \cdot 2!} = \frac{7 \times 6 \times 5 \times 4 \times 3 \times 2 \times 1}{3 \times 2 \times 1 \times 2 \times 1} = 420.$$

Answer: 420

Problem 14 Solution

Let the angular measure of $\overset{\frown}{BD} = x$ degrees. Then the size of $\overset{\frown}{AC} = 180° - 50° - x = 130° - x$. Then $\angle AEC = 20° = \dfrac{x - (130° - x)}{2}$ so solving for x gives $x = 85°$.

Answer: 85

Problem 15 Solution

Using the distance formula we have $AB = \sqrt{25} = 5$, $AC = \sqrt{9} = 3$, and $BC = \sqrt{6}$. Hence using the Law of Cosines,

$$BC^2 = AB^2 + AC^2 - 2 \cdot AB \cdot AC \cdot \cos \angle BAC$$

so

$$6 = 25 + 9 - 2 \cdot 5 \cdot 3 \cdot \cos \angle BAC \Rightarrow \cos \angle BAC = \frac{-28}{-30} = \frac{14}{15}.$$

Therefore $P + Q = 14 + 15 = 29$.

Answer: 29

Problem 16 Solution

We have $\log_{xy}(z) = 2$ so $\log_z(xy) = \log_z(x) + \log_z(y) = 1/2$. We also have $\log_z(x) - \log_z(y) = 1$. Solving this system we get $\log_z(y) = -1/4$, so $\log_y(z) = -4$.

Answer: -4

Problem 17 Solution

All imaginary roots come in conjugates, so we know that $\pm i$ and $-1 \pm i$ are all roots. Hence $p(x)$ either has 0 or 2 real roots. Note $p(x)$ has a positive leading coefficient, so

$$\lim_{x \to +\infty} p(x) = \lim_{x \to -\infty} = +\infty.$$

Combined with the fact that $p(0) = -4$, we see that $p(x)$ has at least one real root. Therefore, we know that $p(x)$ must have 2 real roots. (In fact, the real roots are $-1 \pm \sqrt{3}$, but it is not advised to calculate them to solve this problem.)

Answer: 2

Problem 18 Solution

Note that zeros at the of a number come from powers of 10 as factors. Since $10 = 2 \times 5$ we need to know how many powers of 2 and powers of 5 occur in the prime factorization of 200!. In fact, as there are clearly more powers of 2 than powers of 5, we just need to count how many powers of 5 occur. Multiples of $5^1 = 5, 5, 10, 15, 20, \ldots, 200$, a total of $200 \div 5 = 40$ numbers, each contribute one power of 5. Multiples of $5^2 = 25, 25, 50, \ldots, 200$, a total of $200 \div 25 = 8$ numbers, each contribute one extra power of 5. Lastly, $5^3 = 125$ contributes one more power of 5. Hence in total there are

$$40 + 8 + 1 = 49$$

powers of 5 in the prime factorization of 200!, so there are 49 zeros at the end of 200!.

Answer: 49

Problem 19 Solution

Let r, s be the roots. Hence we know

$$(x-r)(x-s) = x^2 - 101x + k,$$

$r \times s = k$ and $r + s = 101$. Further we are given $r + 4 = 20s$. Hence

$$20s - 4 + s = 101 \Rightarrow 21s = 105 \Rightarrow s = 5 \Rightarrow r = 96.$$

Therefore $k = r \times s = 5 \times 96 = 480$.

Answer: 480

Problem 20 Solution

Consider Jared not qualifying for the next meet. This means that both of his jumps were less than 3 meters. The probability a single jump is less than 3 meters is $\frac{2}{3}$, as he can jump anywhere in the interval $[1,4]$ (with length 3) and the interval of jumps less than 3, $[1,3)$ has length 2. Therefore the probability both jumps are less than 3 meters is

$$\frac{2}{3} \times \frac{2}{3} = \frac{4}{9},$$

that means the probability that he qualifies for the next meet is

$$1 - \frac{4}{9} = \frac{5}{9},$$

so $P + Q = 5 + 9 = 14$.

Answer: 14

2.8   ZIML May 2017 Division H

Below are the solutions from the Division H ZIML Competition held in May 2017.

The problems from the contest are available on p.55.

Problem 1 Solution

Let l, w be the side lengths of the rectangle. Since half the perimeter is 12, we know that $w = 12 - l$. Hence

$$l \times w = 35$$
$$l(12 - l) = 35$$
$$l^2 - 12l + 35 = 0$$
$$(l - 7)(l - 5) = 0,$$

so $l = 7$ or $l = 5$. This implies that $w = 5$ or $w = 7$ so the rectangle is a 7×5 rectangle. Hence the difference between the side lengths is 2.

Answer: 2

Problem 2 Solution

Plugging in the points $(-1, 5)$ and $(3, 5)$ we have $5 = 1 - b + c$ and $5 = 9 + 3b + c$. Solving for b, c gives $b = -2, c = 2$, so the parabola has equation $y = x^2 - 2x + 2$. By symmetry or the formula, the vertex of the parabola (where the minimum occurs) happens when $x = 1$. Plugging in $x = 1$ we get a minimum value of $y = 1 - 2 + 2 = 1$.

Answer: 1

Problem 3 Solution

The units digit of powers of 3 follows the pattern 3, 9, 7, 1, 3, 9, 7, 1, …. This has a cycle of 4. Since 2017 has a remainder of 1 upon division by 4, the units digit is the first term in the pattern:

3.

Answer: 3

Problem 4 Solution
Since the quadratic has integer coefficients, the other root must be $2 - 3i$. Hence

$$4x^2 - 16x + C = 4(x - (2 + 3i))(x - (2 - 3i)).$$

Expanding this we get

$$4x^2 - 16x + C = 4x^2 - 16x + 52,$$

so we see that $C = 52$.

Answer: 52

Problem 5 Solution
Label the resulting pentagon $BEGHI$. We have $AB = BC = 9$, so $BE = 9 - 3 = 6$. Therefore, $FB = 3$ so $CF = 12$. Now note $\triangle CFH$ is a 45-45-90 triangle with hypotenuse 12. Hence its sides are length $12/\sqrt{2}$, so it has area

$$\frac{1}{2} \cdot \frac{12}{\sqrt{2}} \cdot \frac{12}{\sqrt{2}} = 36.$$

Similarly, triangles $\triangle CEG$ and $\triangle FBI$ both have area

$$\frac{1}{2} \cdot 3 \cdot 3 = \frac{9}{2}.$$

Hence, the shaded pentagon has area $36 - 9 = 27$.

Answer: 27

Problem 6 Solution
Arrange the girls first. There are 10 places in total so

$$\frac{10!}{6!} = 10 \cdot 9 \cdot 8 \cdot 7 = 5040$$

ways to decide what place each of the girls finishes in. The boys must occupy the other 6 spots, and as we already know they finish in alphabetical order based on their names, there is only 1 way to arrange the boys in the remaining spots. Hence there are 5040 total outcomes for the race.

Answer: 5040

Problem 7 Solution
Note the to get x^8 when we expand, we must use the x^3 term from the first polynomial and the x^4 term from the last polynomial. We must then have the terms x and 2 from the middle polynomials, in either order. Thus the x^8 terms come from $x^3 \cdot x \cdot 2 \cdot x^4$ or $x^3 \cdot 2 \cdot x \cdot x^4$ so the coefficient is $2 + 2 = 4$.

Answer: 4

Problem 8 Solution
We first find points A and B. It is straightforward to solve the system of equations for A and B. One way to guess and check is as follows. The two numbers must sum to 4, and it is easy to see that $(3, 1), (1, 3)$ both work (as $2^2 = 4$), so $A, B = (1, 3), (3, 1)$.

Since the center of the circle is $C = (1, 1)$, we in fact have that $\triangle ABC$ is a right triangle with both bases of length $3 - 1 = 2$. Hence the area is $\frac{1}{2} \times 2^2 = 2$.

Answer: 2

Problem 9 Solution
The region can be divided into one big sector of $300°$ (5/6 of a circle with radius 4) and two small sectors of $120°$ (each is 1/3 of a circle with radius 1). Thus the area is

$$\frac{5}{6} \cdot 4^2 \pi + \frac{2}{3} \cdot 1^2 \pi = 14\pi,$$

so $K = 14$.

Answer: 14

Problem 10 Solution
Since our list has 6 elements, the median is the average between the 3rd and 4th largest. From this it is clear that $1, 1.5, 2, 2.5, \ldots, 9.5, 10$ are all possible, so a total of 19 medians are possible.

Answer: 19

Problem 11 Solution
Let $\Delta_1 = (-4)^2 - 4(1)(-m) = 4(4+m)$ be the discriminant of the first equation and let $\Delta_2 = (-18)^2 - 4(1)m^2 = 324 - 4m^2$ be the discriminant of the second equation. We want to find all m so that $\Delta_1 < 0$ and $\Delta_2 \geq 0$. Now $\Delta_1 < 0$ when $m < -4$ and $\Delta_2 \geq 0$ when $m^2 \leq 81$ or $-9 \leq m \leq 9$. The overlap of these two sets is $-9 \leq m < -4$. Hence $m = -9, -8, -7, -6, -5$ are the only integer values, so there are 5 such m.

Answer: 5

Problem 12 Solution
Use the Law of Cosines to get $3^2 = 3^2 + 4^2 - 2 \cdot 3 \cdot 4 \cos(\angle CBD)$ and solve to get $\cos(\angle CBD) = \frac{2}{3}$. Hence using the Law of Cosines again we have $(AD)^2 = 9^2 + 4^2 - 2 \cdot 9 \cdot 4 \cdot \cos(\angle CBD) = 81 + 16 - 72 \cdot \frac{2}{3} = 49$ so $AD = 7$.

Answer: 7

Problem 13 Solution

We have (using change of base)

$$\log_3(x) + \log_9(x) + \log_{27}(x) = \log_3(x) + \frac{\log_3(x)}{\log_3(9)} + \frac{\log_3(x)}{\log_3(27)}$$
$$= \log_3(x) + \frac{\log_3(x)}{2} + \frac{\log_3(x)}{3}$$
$$= \frac{11\log_3(x)}{6}.$$

Hence we need $\log_3(x) = 4$ so $x = 3^4 = 81$.

Answer: 81

Problem 14 Solution

First note $2800 = 2^4 \times 5^2 \times 7$. Since A has 9 divisors, it is a square number of the form $p^2 q^2$ (p and q are primes) or the form p^8, based on the formula to count divisors. From the factorization of 2800, A can only be $2^2 \times 5^2 = 100$.

Based on the definition of LCM, B has to have factor $2^4 \times 7$. Notice that $2^4 \times 7$ has exactly 10 factors, so we conclude that B is just $2^4 \times 7$ which equals 112.

Therefore, $A = 100, B = 112$, so the larger is 112.

Answer: 112

Problem 15 Solution

It is easy to calculate the ℓ_3 intersects ℓ_1, ℓ_2 at $(4,2), (2,4)$ and the midpoint of these is $(3,3)$. Note that $\ell_1 \| \ell_2$, so if we have a circle at $(3,3)$ that is tangent to ℓ_1 it is also tangent to ℓ_2. Therefore, the largest circle must have center $(3,3)$ and radius $\sqrt{2}$. The square of the radius is 2.

Answer: 2

Problem 16 Solution

Let $z = 2x$. Note $\sin(z - 90°) = -\sin(90° - z) = -\cos(z)$. Solving $\cos(z) = -\sqrt{2}/2$ we have $z = 135° + 360°k$ or $z = 225° + 360°k$ for all integers k. Hence $x = 67.5° + 180°k$ or $x = 112.5° + 180°k$ for all integers k. Hence

$$x = 67.5°, 247.5°, 112.5°, 292.5$$

are all the solutions from $0°$ to $360°$. The sum of these is $720°$.

Answer: 720

Problem 17 Solution
Checking possible roots from the Rational Root Theorem, we see $1, 2$ are roots, so we can divide by $(x - 1)(x - 2) = x^2 - 3x + 2$. Using long division this gives us

$$(x^4 - 3x^3 - x^2 + 9x - 6) \div (x^2 - 3x + 2) = x^2 - 3,$$

so the other roots are $\pm\sqrt{3}$. Hence $A + B = 0 + 3 = 3$.

Answer: 3

Problem 18 Solution
The volume of a cone is

$$\frac{1}{3} \times \pi R^2 \times H$$

where R is the radius of the base and H is the height. We are given that the height is $H = 10$ and we can calculate the radius using the distance formula,

$$R = \sqrt{(1 - (-2))^2 + (1 - 3)^2 + (1 - (-5))^2}$$
$$= \sqrt{9 + 4 + 36} = \sqrt{49} = 7.$$

Hence the volume is

$$\frac{1}{3} \times \pi 7^2 \times 10 = \frac{490\pi}{3},$$

so $C = \dfrac{490}{3} \approx 163.33$.

Answer: 163.33

Problem 19 Solution

The full interval has length 10. We have $|x-1| > 2$ if $x > 3$ or $x < -1$. Hence $|x-1| > 2$ if x is in the interval $[3,5]$ (of length 2) or the interval $[-5,-1]$ of length 4. Hence the probability (the ratio of lengths) is

$$\frac{2+4}{10} = \frac{6}{10} = 60\%,$$

and $P = 60$ is the answer.

Answer: 60

Problem 20 Solution

Rewrite the equation to $\lfloor x \rfloor + \{x\} + \{x\} = 2\lfloor x \rfloor$, so $2\{x\} = \lfloor x \rfloor$. Since $0 \le \{x\} < 1$, and $\lfloor x \rfloor$ is an integer, we get $\{x\} = 1/2$, and $\lfloor x \rfloor = 1$. Therefore $x = 3/2 = 1.5$.

Answer: 1.5

2.9 ZIML June 2017 Division H

Below are the solutions from the Division H ZIML Competition held in June 2017.
The problems from the contest are available on p.61.

Problem 1 Solution
The sum of all the widths is equal to

$$12 \times 5.2 + 8 \times 4.7 = 100$$

So the mean width of the 12 iPads and 8 Kindles is

$$100 \div 20 = 5.$$

Answer: 5

Problem 2 Solution
$\dfrac{1000}{13} \approx 76.9$, so there are 76 multiples of 76 under 1000.
$\dfrac{1000}{17} \approx 58.8$, so there are 58 multiples of 17 under 1000.
The overlap of these two groups are the multiples of $13 \cdot 17 = 221$.
Since

$$\frac{1000}{221} \approx 4.5$$

there are 4 overlapped numbers. Therefore the answer is

$$76 + 58 - 4 = 130.$$

Answer: 130

Problem 3 Solution
We first have using adjacent angles

$$\angle DLN = 180° - 135° = 45°.$$

Since $\overline{AB} \parallel \overline{CD}$,

$$\angle NDL = \angle AJG = 30°$$

as they are corresponding angles. Therefore, as the angles in triangle $\triangle DNL$ add up to $180°$,

$$\angle LND = 180° - 45° - 30° = 105°.$$

Answer: 105

Problem 4 Solution

Let x, y, z denote the number of hours Bill, Claire, Drew drove. Hence from the information given we know $x + y + z = 10$, $40x + 50y + 60z = 485$, and $40x = 60z$. We want to solve for y, so we can substitute the third equation into the first two (for better numbers, multiply the first equation by 40 first), giving the system of equations

$$\begin{cases} 40y + 100z &= 400, \\ 50y + 120z &= 485. \end{cases}$$

Multiplying the first equation by $\dfrac{6}{5}$ we get $48y + 120z = 480$ so subtracting we have $2y = 5$ so $y = 2.5$. Hence Claire drives for 2.5 hours.

Answer: 2.5

Problem 5 Solution

We subtract the areas of the unshaded triangles from the total area (which is 25). The triangle above $\overline{AD}$ is half of a parallelogram of area 6, so has area 3. The triangle to the right of $\overline{AB}$ is half a triangle of area 4, so has area 2. Lastly, the triangle below $\overline{BC}$ is half a parallelogram of area 16, hence has area 8. Hence the area of $ABCD$ is $25 - 3 - 2 - 8 = 12$.

Answer: 12

Problem 6 Solution

Note the vertex of f occurs at $-(-6)/2 = 3$ (the vertex is $(3,3)$, so f has an inverse for $x \geq 3$ and thus $m = 3$.

Answer: 3

Problem 7 Solution

Completing the square we have the parabola is $y - 2 = 2(x - 2)^2$ hence it has a vertex $(2,2)$ and opens upwards. For the circle we have $(x - 2)^2 + y^2 = 4$ so it is a circle of radius 2 with center $(2,0)$. From here it is clear that they only intersect at the vertex of the parabola: $(2,2)$. Hence they intersect only once.

Answer: 1

Problem 8 Solution

Note $\overline{EF}, \overline{GH}$ divide $ABCD$ into four parallelograms, whose areas are proportional as they share sides or heights. That is

$$\frac{[AEPG]}{[GPFD]} = \frac{[EBHP]}{[PHCF]}$$

so $[AEPG] = 10 \times 16 \div 8 = 20$. Hence the total area of $ABCD$ is $10 + 8 + 16 + 20 = 54$.

Answer: 54

Problem 9 Solution

In order to trade 3 books, the first student must select 3 books from the 6 books he has. The second student must select 3 books from the 8 books. Since the order of the books does not matter, the first student can do this in

$$\binom{6}{3} = 20$$

ways and the second student can do this in

$$\binom{8}{3} = 56$$

ways. Therefore, the total number of ways that this can be done is:

$$\binom{6}{3} \times \binom{8}{3} = 20 \times 56 = 1120.$$

Answer: 1120

Problem 10 Solution

The shortest path goes through the midpoint of one of the edges, as shown below. (The path is shortest because it is a straight line when the faces are unfolded.)

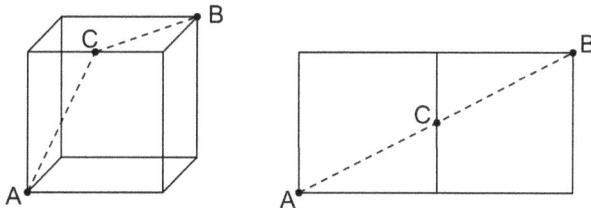

Using the Pythagorean theorem we have that

$$AC = BC = \sqrt{2^2 + 1^2} = \sqrt{5},$$

so the shortest path has length $2\sqrt{5}$. Hence $P + Q = 2 + 5 = 7$.

Answer: 7

Problem 11 Solution

Let $y = x^2$ so we have

$$\frac{1}{y} - \frac{1}{y+3} = \frac{3}{4}.$$

Clearing denominators gives $4y + 12 - 4y = 3y^2 + 9y$ so

$$3y^2 + 9y - 12 = 3(y^2 + 3y - 4) = 3(y+4)(y-1) = 0$$

and hence $y = 1, -4$. Thus $x^2 = 1$ giving the above mentioned solutions $x = \pm 1$ or $x^2 = -4$ giving two more solutions $x = \pm 2i$. Hence $K = 2$.

Answer: 2

Problem 12 Solution

Using the Law of Cosines we have

$$BC^2 = 3^2 + (2\sqrt{2})^2 - 2 \times 3 \times 2\sqrt{2}\cos(45°)$$

$$= 9 + 8 - 12\sqrt{2} \times \frac{\sqrt{2}}{2} = 5,$$

so $BC = \sqrt{5}$. Using the Law of Sines

$$\frac{\sin(A)}{BC} = \frac{\sin(C)}{AB} \Rightarrow \frac{\sqrt{2}}{2\sqrt{5}} = \frac{\sin(C)}{3},$$

and thus

$$\sin(C) = \frac{3}{\sqrt{10}} = \frac{3\sqrt{10}}{10}.$$

Using the Law of Cosines we calculate

$$3^2 = (2\sqrt{2})^2 + (\sqrt{5})^2 - 2 \times 2\sqrt{2} \times \sqrt{5}\cos(C)$$

and thus

$$\cos(C) = \frac{-4}{-4\sqrt{10}} = \frac{\sqrt{10}}{10}.$$

Hence $\tan(C) = 3$. (Note using the identity $\sin^2(C) + \cos^2(C) = 1$ we can determine that $\cos(C) = \pm\frac{\sqrt{10}}{10}$, but we still need to justify that $\cos(C)$ is positive, which is why we use the Law of Cosines again above.)

Answer: 3

Problem 13 Solution
First note that $P(6)$ is missing. Since all probabilities must add up to 1, we have

$$P(6) = 1 - 0.2 - 0.1 - 0.2 - 0.1 - 0.15 = 0.25.$$

Hence the probability of getting an even number is

$$P(2) + P(4) + P(6) = 0.1 + 0.1 + 0.25 = 0.45,$$

so $K = 45$.

Answer: 45

Problem 14 Solution
Note that the area of the intersection of the circles inside the square can be determined by finding the area of two quarter circles and subtracting the area of the square. Therefore, the area of the intersection of the circles is

$$\frac{1}{4}\pi \times 2^2 + \frac{1}{4}\pi \times 2^2 - 2^2 = 2\pi - 4.$$

As $\pi \approx 3.14$ our answer is approximately $2 \times 3.14 - 4 = 6.28 - 4 \approx 2.28$.

Answer: 2.28

Problem 15 Solution
Rewrite the equation as

$$\frac{1}{\log_2(x)} + \log_2(x) = \frac{5}{2}.$$

Substitute $z = \log_2(x)$ and solve $\dfrac{1}{z} + z = \dfrac{5}{2}$ to get $z = 2$ or $z = 1/2$. This gives $x = 4$ or $x = \sqrt{2}$. Hence the sum is $4 + \sqrt{2} \approx 4 + 1.414 \approx 5.4$ rounded to the nearest tenth.

Answer: 5.4

Problem 16 Solution

We have that $140 = 2^2 \cdot 5 \cdot 7$. Therefore x is either 2, 5, or 7.

First assume $x = 7$, so we have $(y + 7)(y + z) = 20$. As $y, z \geq 2$, we have $(y + 7)(y + z) \geq 9 \cdot 4 = 36$ so this is impossible.

Next assume $x = 5$, so we have $(y + 5)(y + z) = 28$. As $y, z \geq 2$, the first term must be at least 7 and the second term must be at least 4, leaving only $28 = 7 \cdot 4$ which implies $y = 2$ and $z = 2$.

Lastly assume $x = 2$, so we have $(y + 2)(y + z) = 70$. As $y, z \geq 2$, we look at the factor pairs: $(14, 5)$, $(10, 7)$, $(7, 10)$, and $(5, 14)$. Of these, only $(7, 10)$ and $(5, 14)$ work, leading to pairs $(5, 5)$ and $(3, 11)$.

Hence there are three triples in total: $(5, 2, 2)$, $(2, 5, 5)$, and $(2, 3, 11)$, only $(2, 3, 11)$ has all three primes different, so our answer is $2 + 3 + 11 = 16$.

Answer: 16

Problem 17 Solution

First note $AB = 3$. Then using the distance formula we have

$$BC = \sqrt{0^2 + (2\sqrt{2})^2 + (2\sqrt{2})^2} = \sqrt{8 + 8} = 4,$$

and

$$AC = \sqrt{3^2 + (2\sqrt{2})^2 + (2\sqrt{2})^2} = \sqrt{9 + 8 + 8} = 5.$$

Hence $\triangle ABC$ is a right triangle with base and height 3 and 4, so its area is $\dfrac{1}{2} \times 3 \times 4 = 6$.

Answer: 6

Problem 18 Solution

The number is a multiple of $72 = 2^3 \times 3^2$. We have that $15 = 15 \times 1 = 5 \times 3$ so since the number we want has 15 factors, it is either in the form p^{14} or $p^4 \times q^2$ for primes p, q. The number must have two prime factors, so it must be of the form $p^4 \times q^2$. Further, it must have a factor of 2^3, so we must have $p = 2, q = 3$ so the number is $2^4 \times 3^2 = 144$. Since this is the only possibility, it is the largest, so 144 is the answer.

Answer: 144

Problem 19 Solution

Using the double angle formula we can rewrite everything in terms of $\sin(x), \cos(x)$:

$$2\sin(x)\cos(x) = \frac{\sin(x)}{\cos(x)}.$$

Hence either $\sin(x) = 0$, so $x = 0°, 180°$. If $\sin(x) \neq 0$ we can cancel $\sin(x)$ and solve for $\cos(x)$ to get

$$\cos^2(x) = \pm\frac{1}{2} \text{ so } \cos(x) = \pm\frac{\sqrt{2}}{2}.$$

This gives additional solutions $x = 45°, 135°, 225°, 315°$. Therefore in all there are $2 + 4 = 6$ solutions with $0° \leq x < 360°$.

Answer: 6

Problem 20 Solution

First note

$$\left(\frac{\sqrt{2}}{2}(-1+i)\right)^2 = \frac{1}{2}(1-2i+i^2) = -i.$$

Hence

$$\left(\frac{\sqrt{2}}{2}(-1+i)\right)^{100} = (-i)^{50} = (-1)^{25} = -1$$

so $B - A = 0 - (-1) = 1$.

Answer: 1

3. Appendix

3.1 Division H Topics Covered

Algebra

- Students should be comfortable with ratios, proportions, and their applications to problems involving work and motion, but these problems are not a main focus at this level
- Radicals, Exponents, and Logarithms: Simplest Radical Form for Roots, Laws of Exponents, Laws of Logarithms including change of base
- Complex Numbers: Arithmetic Operations and writing in rectangular form
- Factoring Tricks: Sums and differences of squares, cubes, etc., Expanding $(x+y)^n$ using Pascal's triangle
- Solving Equations: Linear Equations, Quadratic Equations, Systems of Equations, Substitutions to rewrite higher degree equations as quadratics, Radicals, Absolute Values
- Quadratics: Graphing and Vertex Form, Maxima and Minima, Quadratic Formula, Discriminant

- Polynomials: Polynomial Long Division, Remainder and Factor Theorem, Rational Root Theorem

Geometry

- As a general rule students should be comfortable using algebraic techniques (linear equations, quadratic equations, systems of equations, etc.) as tools for applying the geometric concepts listed below
- Angles in Parallel Lines (corresponding angles, alternating interior/exterior angles, same-side interior/exterior angles, etc.)
- Analytic Geometry: Equations of Lines, Parabolas, and Circles, Distance Formula, Midpoint Formula, Geometric Interpretation of Slope and Angles
- Triangles: Congruence and Similarity, Pythagorean theorem, Ratios of Sides for triangles with angles of 45, 45, 90 or 30, 60, 90
- Trigonometry: General understanding of sine, cosine, tangent, and their cofunctions, Law of Sines and Cosines, Trigonometric Identities for double angles, sums/differences, etc.
- Centers in Triangles: Definitions of altitudes, medians, angle bisectors, perpendicular bisectors
- Interior and Exterior Angles of Polygons, including the sum of all these angles, each angle if the polygon is regular, etc.
- Areas and Perimeters of basic shapes such as triangles, rectangles, parallelograms, trapezoids, and circles, Heron's formula and formulas using inradius or circumradius for triangles
- Geometric Reasoning with Areas: Congruent shapes have the same area, Similar triangles have a ratio of areas that is the square of the ratio of their sides, Triangles with the same height have a ratio of their areas equal to the ratio of their bases, etc., Using multiple expressions of area to solve for unknowns

- Circles: Arc Length, Sector Area, Definitions for Tangent Lines and Tangent Circles, Inscribed Angles, Angles formed by intersecting chords
- Solid Geometry: Surface Area and Volume for Spheres, Prisms, Pyramids, and Cones, Reasoning for more general solids, such as combining the solids listed above or pieces of solids when cut by a plane, etc.

Counting and Probability

- Fundamental Rules: Sum and Product Rules, Permutations and Combinations
- Counting Methods: Complementary counting, Grouping objects that must be together, Inserting objects that must be apart into spaces between objects, etc.
- Sequences: Arithmetic and Geometric Sequences and Series, Finding and understanding patterns and recursive definitions for general sequences
- Probability and Sets: Definitions for event, sample space, complement, intersection, and union, Understanding the use of Venn Diagrams
- Probability: In finite sample spaces as a ratio of the number of outcomes, In geometric sample spaces as a ratio of lengths, areas, or volumes, Axioms of Probability, Independence, Conditional Probability, Law of Total Probability

Number Theory

- Fundamental Definitions: Prime numbers, factors/divisors, multiples, least common multiple (LCM), greatest common factor/divisor (GCF or GCD), perfect squares/cubes/etc.
- Divisibility Rules for numbers such as 2, 3, 4, 5, 8, 9, 10, 11, and how to combine the rules for numbers such as 6, 22, etc.

- (Unique) Prime Factorization and how to use the prime factorization to find the number of factors, to test whether a number is a perfect square/cube/etc, to find the LCM or GCD.
- Factoring Tricks: Factors come in pairs, perfect squares have an odd number of factors, etc.
- Remainders: Patterns for finding remainders, for example units digits or last two digits

3.2 Glossary of Common Math Terms

Acute Angle An angle less than $90°$.

Altitude of a Triangle A line segment connecting a vertex of a triangle to the opposite side forming a right angle. Also called the height of a triangle.

Angle A figure formed by two rays sharing a common vertex. Often measured in degrees.

Angle Bisector A line dividing an angle into two equal halves.

Arc The curve of a circle connecting two points.

Area The amount of space a region takes up. Often denoted using square brackets: area of $\triangle ABC = [ABC]$.

Arithmetic Sequence A sequence where the difference between one term and the next is constant.

Average See Mean.

Base of a Triangle One side of a triangle, often used when the altitude is drawn from the opposite side to this base.

Binomial Coefficient The symbol $\binom{n}{k} = \dfrac{n!}{k!(n-k)!}$.

Chord A line segment connecting two points on the outside of a circle.

Circle A round shape consisting of points that all have the same distance (called the radius) from the center of the circle.

Circumference The perimeter of a circle.

Circumscribe To draw a shape outside another shape so that the boundaries touch.

Coefficient The number being multiplied by a variable or power of a variable. For example, the coefficient of x^3 in $5x^5 + 4x^3 + 2x$ is 4.

Complement In probability, the complement of a set is all elements outside the set.

Composite Number A number that is not prime.

Congruent Two shapes or figures that are exactly the same.

Cube A solid figure formed by 6 congruent squares that all meet at right angles.

Deck of Cards A standard deck of cards has 52 cards. There are 4 suits (clubs, diamonds, hearts, and spades) with each suit having cards of 13 ranks (A (ace), $2, 3, \ldots, 10$, J (jack), Q (queen), and K (king)).

Degree of a Polynomial The highest power of a variable in the polynomial. For example, the degree of $2x^3 - 5x^6 + 2$ is 6.

Denominator The bottom number in a fraction.

Diagonal A line segment connecting two vertices of a shape or solid that is not an edge of the shape or solid.

Diameter A chord passing through the center of a circle. The diameter has length that is twice the radius.

Die or Dice A standard die (plural is dice) has 6 sides. Each of the 6 sides has the same chance when the die is rolled.

Digit One of $0, 1, 2, \ldots, 9$ used when writing a number.

Discriminant The expression $b^2 - 4ac$ for a quadratic equation $ax^2 + bx + c = 0$.

Distinguishable Objects Objects that are different.

Divisible A number is divisible by another number if there is no remainder when the first number is divided by the second. For example, 35 is divisible by 7.

Divisor A number that evenly divides another number. For example, 6 is a divisor of 48. Also called a factor.

Edge A line segment connecting two vertices on the outside of a shape or solid.

Equally Likely Having the same chance of occurring.

Equiangular Polygon A shape with all equal angles.

Equilateral Polygon A shape with all equal sides.

Equilateral Triangle A regular triangle, one with three equal sides and three equal angles.

Even Number A number divisible by 2.

Exponent The number another number is raised to for powers. For example, in a to the power of b (a^b), the exponent is b.

Face The shape or polygon on the outside of a solid region.

Factor of a Number A number that evenly divides another number. For example, 6 is a factor of 48. Also called a divisor.

Factorial The symbol ! where $n! = n \times (n-1) \times (n-2) \cdots \times 1$.

Fraction An expression of a quotient. For example, $\frac{1}{2}$ or $\frac{9}{7}$.

Function A function is a rule that associates exactly one output with every input. Often described using an equation.

Geometric Sequence A sequence where the ratio between one term and the next is constant.

Greatest Common Divisor (GCD) The largest number that is a divisor/factor of two or more numbers.

Greatest Common Factor (GCF) See Greatest Common Divisor.

Indistinguishable Objects Objects that are the same.

Inscribe To draw a shape inside another shape so that the boundaries touch.

Intersecting Lines or curves that cross each other.

Intersection of Two Sets The set of objects that are in both of the two sets. Denoted using $\cap$. For example, $\{2,3\} \cap \{3,4,5\} = \{3\}$.

Isosceles Triangle A triangle with two equal sides and two equal angles.

Least Common Multiple (LCM) The smallest number that is a multiple of two or more numbers.

Mean The sum of the numbers in a list divided by the how many numbers occur in the list. Also called the average.

Median The number in the middle of a list when the list is arranged in increasing order.

Median of a Triangle A line connecting a vertex in a triangle to the midpoint of the opposite side.

Midpoint The point in the middle of a line segment.

Mode The number or numbers occurring most often in a list of numbers.

Multiple A number that is an integer times another number. For example, 72 is a multiple of 8.

Numerator The top number in a fraction.

Obtuse Angle An angle between $90°$ and $180°$.

Odd Number A number not divisible by 2.

Parallel Lines Lines that do not intersect.

Perfect Cube A number that is another number cubed. For example, $64 = 4^3$ is a perfect cube.

Perfect Square A number that is another number squared. For example, $64 = 8^2$ is a perfect square.

Perimeter The length/distance around the outside of a shape.

Perpendicular Bisector A line perpendicular to and passing through the midpoint of a line segment.

Pi (π) A number used often in geometry. $\pi = 3.1415926\ldots \approx 3.14 \approx \frac{22}{7}$.

Polygon A shape formed by connected line segments.

Polynomial A function that is made of adding multiples of powers of a variable. For example, $f(x) = x^4 + 3x^2 + 2x - 3$.

Prime Factorization The expression of a number as the product of all its prime factors. For example, 24 has prime factorization $2 \times 2 \times 2 \times 3 = 2^3 \times 3$.

Prime Number A number whose only factors are one and itself.

Proportional Ratios Ratios that have equal values when expressed in fraction form. For example, $2 : 3$ is proportional to $8 : 12$.

Quadratic A polynomial with degree 2. Often written in the form $ax^2 + bx + c$.

Quadrilateral A shape with four sides.

Quotient The integer quantity when dividing one number by another. For example, the quotient of $38 \div 5$ is 7 as $38 = 7 \times 5 + 3$.

Radius of a Circle The distance from the center of the circle to any point on the outside of the circle.

Randomly Chosen for a group of objects. Unless specified, the chance of choosing each object is the same as any other object.

Rank of a Card See Deck of Cards.

Ratio A relation depicting the relation between two quantities. For example $2 : 3$ or $\frac{2}{3}$ denotes that for every 3 of the second quantity there are 2 of the first quantity.

Rational Number A number that can be written as a fraction.

Reciprocal One divided by the number. For example, the reciprocal of 7 is $\frac{1}{7}$.

Rectangle A quadrilateral with four right angles (an equiangular quadrilateral).

Rectangular Form (of a complex number) A complex number written in the form $a + b \cdot i$ for real numbers a and b.

Regular Polygon A polygon with all equal sides and all equal angles (equilateral and equiangular).

Remainder The quantity left over when one integer is divided by another. For example, the remainder of $38 \div 5$ is 3 as $38 = 7 \times 5 + 3$.

Rhombus A quadrilateral with four equal sides (an equilateral quadrilateral).

Right Angle A $90°$ angle.

Right Triangle A triangle containing a right angle.

Root of a Function A value of x such that the function evaluates to zero. For example, $x = 2$ is a root of the function $f(x) = x^2 - 4$.

Sample Space In probability, the sample space is the set of all outcomes for an experiment.

Scalene Triangle A triangle with three unequal sides and three unequal angles.

Sector The region formed by an arc and the two radii connecting the ends of the arc to the center of the circle.

Sequence An ordered list of numbers.

Set An unordered collection or group of objects without repeated elements. Denoted using curly brackets. For example, $\{1,2,3,4\}$ is the set containing the integers $1,\ldots,4$.

Similar Shapes or solids that have the same angles and sides that share a common ratio.

Simplest Radical Form An expression containing a radical such that the number inside the radical is an integer that has no perfect squares.

Sphere A round solid consisting of points that all have the same distance (called the radius) from the center of the sphere.

Square A shape with four equal sides and four equal angles (a regular quadrilateral).

Subset A set of objects that is contained inside a larger set of objects. Denoted using $\subseteq$. For example $\{2,3\} \subseteq \{1,2,3,4\}$.

Suit of a Card See Deck of Cards.

Surface Area The total area of all the faces of a solid.

Tangent Line A line touching a shape or curve at exactly one point.

Trapezoid A quadrilateral with one pair of parallel sides.

Triangle A shape with three sides.

Union of Two Sets The set of objects that are in one or both of the two sets. Denoted using $\cup$. For example, $\{2,3\} \cup \{3,4,5\} = \{2,3,4,5\}$.

Venn Diagram A diagram with circles used to understand the relationship between overlapping sets.

Vertex The intersection of line segments, especially the intersection of sides or edges in a shape or solid.

Volume The amount of space a solid region takes up.

With Replacement When choosing objects with replacement, a chosen object is returned to the others allowing it to be chosen more than once.

3.3   ZIML Answers

ZIML October 2016 Division H

Problem 1:	6	Problem 11:	-12
Problem 2:	7	Problem 12:	-18
Problem 3:	6.35	Problem 13:	216
Problem 4:	-15	Problem 14:	3
Problem 5:	4	Problem 15:	5
Problem 6:	15	Problem 16:	4
Problem 7:	5	Problem 17:	16
Problem 8:	97	Problem 18:	35
Problem 9:	2	Problem 19:	5
Problem 10:	10	Problem 20:	7

ZIML November 2016 Division H

Problem 1: -6 Problem 11:   8

Problem 2:   13   Problem 12:   3

Problem 3:   86   Problem 13:   8

Problem 4:   28   Problem 14:   368

Problem 5:   60   Problem 15:   2

Problem 6:   90   Problem 16: -0.8

Problem 7:   10   Problem 17:   128

Problem 8:   102   Problem 18:   20

Problem 9:   6   Problem 19:   2

Problem 10:   900   Problem 20:   7

ZIML December 2016 Division H

Problem 1:   8   Problem 11:   4

Problem 2:   18   Problem 12:   120

Problem 3:   3600   Problem 13:   12

Problem 4:   14   Problem 14:   0.75

Problem 5:   4   Problem 15:   256

Problem 6:   48   Problem 16:   9

Problem 7:   32   Problem 17: -1

Problem 8:   16   Problem 18:   13

Problem 9: -2 Problem 19:   18

Problem 10:   23   Problem 20:   1

ZIML January 2017 Division H

Problem 1:	67	Problem 11:	-4
Problem 2:	107	Problem 12:	36
Problem 3:	21	Problem 13:	9
Problem 4:	46	Problem 14:	11
Problem 5:	15	Problem 15:	25
Problem 6:	5040	Problem 16:	2
Problem 7:	40	Problem 17:	25
Problem 8:	9	Problem 18:	46
Problem 9:	4	Problem 19:	2701
Problem 10:	54	Problem 20:	0.25

ZIML February 2017 Division H

Problem 1:   1   Problem 11:   239

Problem 2:   46   Problem 12:   31

Problem 3:   10   Problem 13:   28

Problem 4:   1440   Problem 14:   40

Problem 5:   9.2   Problem 15:   −69

Problem 6:   8   Problem 16:   150

Problem 7:   5   Problem 17:   625

Problem 8:   200   Problem 18:   48

Problem 9:   6   Problem 19:   60

Problem 10:   120   Problem 20:   0

ZIML March 2017 Division H

Problem 1:	1.5	Problem 11:	0.67
Problem 2:	22	Problem 12:	135
Problem 3:	6	Problem 13:	5
Problem 4:	7	Problem 14:	132
Problem 5:	2	Problem 15:	8
Problem 6:	3	Problem 16:	3
Problem 7:	101	Problem 17:	2019
Problem 8:	-25	Problem 18:	210
Problem 9:	46	Problem 19:	87.5
Problem 10:	11	Problem 20:	225

ZIML April 2017 Division H

Problem 1: 2.7

Problem 2: 75

Problem 3: 50

Problem 4: 51

Problem 5: 6

Problem 6: 2

Problem 7: 10.33

Problem 8: 3

Problem 9: 4

Problem 10: 7.1

Problem 11: 14

Problem 12: -48

Problem 13: 420

Problem 14: 85

Problem 15: 29

Problem 16: -4

Problem 17: 2

Problem 18: 49

Problem 19: 480

Problem 20: 14

ZIML May 2017 Division H

Problem 1:	2	Problem 11:	5
Problem 2:	1	Problem 12:	7
Problem 3:	3	Problem 13:	81
Problem 4:	52	Problem 14:	112
Problem 5:	27	Problem 15:	2
Problem 6:	5040	Problem 16:	720
Problem 7:	4	Problem 17:	3
Problem 8:	2	Problem 18:	163.33
Problem 9:	14	Problem 19:	60
Problem 10:	19	Problem 20:	1.5

ZIML June 2017 Division H

Problem 1:   5   Problem 11:   2

Problem 2:   130   Problem 12:   3

Problem 3:   105   Problem 13:   45

Problem 4:   2.5   Problem 14:   2.28

Problem 5:   12   Problem 15:   5.4

Problem 6:   3   Problem 16:   16

Problem 7:   1   Problem 17:   6

Problem 8:   54   Problem 18:   144

Problem 9:   1120   Problem 19:   6

Problem 10:   7   Problem 20:   1

www.ingramcontent.com/pod-product-compliance
Lightning Source LLC
Chambersburg PA
CBHW071856200326
41519CB00016B/4406